Civil Engineering Contracts

Second edition

Civil Engineering Contracts
Practice and Procedure

Second Edition

Charles K. Haswell, BSc(Eng)Lond F Eng, FICE, FIStructE, MConsE
Senior Partner, Charles Haswell & Partners

Douglas S. de Silva, CEng, FICE, FBIM
Partner, Charles Haswell & Partners

Butterworth
London Boston Singapore Sydney Toronto Wellington

First published 1989   1072125 8

© Butterworth & Co. (Publishers) Ltd, 1989

British Library Cataloguing in Publication Data

Haswell, Charles, K.
  Civil engineering contrast : practice and procedure.
  1. England. Civil engineering. Contracts. Law
  I. Title   II. De silva, Douglas S.
  344.203′78624

  ISBN 0-408-03201-4

Library of Congress Cataloging in Publication Data

Haswell, Charles K.
  Civil engineering contracts : practice and procedure /
Charles K. Haswell, Douglas S. de Silva. – 2nd ed.
  228 p. 21.6 cm.
  Bibliography: p. 224
  Includes index.
  ISBN 0-408-03201-4 :
  1. Civil engineering–Contracts and specifications–
  Great Britain.
  I. De Silva, Douglas S.   II. Title.
  KD1641.H37   1989
  343.41′078624--dc19   89-951
  [344.10378624]   CIP

Photoset by Butterworths Litho Preparation Department
Printed in Great Britain at the University Press, Cambridge

26.4.94

Foreword to the first edition

By Sir Edward Singleton

In his Presidential Address to the Institution of Civil Engineers in 1963, Sir Harold Harding referred to two types of members of the profession – the gum-boot party and the egg-head party. He went on to say he belonged to the party of the extreme centre.

It was at about the same time that Sir Harold introduced me to Charles Haswell, and I quickly recognized another of the same type: an engineer with any amount of technical and theoretical ability, but one who would not make his mind up until he had been down in the tunnel or up on the dam to find out for himself.

It was for this reason that I was delighted to hear that Charles Haswell was to write a book about these subjects that he knows so well, and was to be joined in this by another engineer as capable and well qualified as Douglas de Silva; and that the book would deal with the elements of civil engineering contracts and their administration.

One can easily fill one's bookshelf with large and impressive books on the technical side of engineering and on the various conditions of contract, and whole chapters have been written about the niceties of contract interpretation. What many of us would like to have had, both inside and outside the profession, when we first became involved in this specialized world of building and civil engineering, is a simple book which explains how the construction projects are put together and even what some of the specialized language means.

I believe that this book will achieve that object, as well as being a basic textbook for the practising engineer who becomes involved with contract administration. I wish it every success.

v

Preface to the second edition

Some seven years have passed since this book was first published. During this time the change in procedures for the implementation of projects has been rapid. Attitudes of employers in particular have also changed, and they now relate to methods of financing projects, types of contract and the selection of professional advisers to act in the role of engineer or as design consulting engineers. During this period some new standard forms of contract have been published and some revised, as have procedures in arbitration.

It follows that an update of this book was called for. In this edition we have continued to maintain the traditional approach on contract procedures adopted on civil engineering projects and have taken into account the new developments in this field as far as it is practicable in a book of this size. The new coverage includes comments on the second edition of the Civil Engineering Standard Method of Measurement (CESMM2), the fourth edition of the FIDIC Conditions of Contract for Works of Civil Engineering Construction and the latest Institution of Civil Engineers' Arbitration Procedure. Reference has also been made to new forms of contract that have been published and almost every chapter of the earlier edition has been revised and updated.

The authors hope that this volume will continue to assist civil engineers in general and those younger engineers who are aspiring to obtain professional qualifications.

Charles K. Haswell
Douglas S. de Silva

Preface to the first edition

During the past few decades many aspects of civil engineering projects have become more complicated. Advances in design, in technology and in construction techniques have largely been responsible for this change. This has had a corresponding effect in the contract field to which are added modern-day restraints such as cash flow problems.

The two parties to any contract have naturally different objectives: the employer seeks value for money; the contractor has to ensure profitability. The engineer as the independent professional civil engineer has to act as the technical expert and at the same time as administrator and arbiter, as it were, for the execution of the contract. For the reasons given, the engineer's task has tended to become more onerous as time passes; the changes in modern conditions of contract, lessening both the contract risk (q.v.) and the power of the engineer have also increased the complexity and difficulty of the engineer's work; moreover, his level of responsibility remains sensibly similar over this period.

The foregoing has led to an ever-increasing need for further education in contract matters. In the UK the Institution of Civil Engineers has recognized the requirement with an examination in Civil Engineering Law and Contract Procedure: the object of the qualification is to equip the better those in the profession as their responsibilities increase, as well as those wishing to qualify later as arbitrators. The authors hope this volume will assist civil engineers in general and those preparing for the additional qualification. Much of the text has been taken from lectures given on contract procedure for the foregoing qualification.

Charles K. Haswell
Douglas S. de Silva

Acknowledgements

The authors wish to thank not a few readers of the first edition for their helpful comments which have been incorporated as appropriate. They would, in particular, like to record their thanks to Sir Edward Singleton in general and Mr Jonathan Blackmore on insurance matters for their assistance in producing this new edition of the book.

They would also like to acknowledge the kind permission given by the following organizations to reproduce material published by them:

The Institution of Civil Engineers for the inclusion of:

1. Appendix 2. Generic case for cost/reimbursement contract with twin targets from ICE Paper 7650, 'Tunnel under the Severn Wye Estuaries', by C. K. Haswell, *Proceedings ICE*, August 1973;
2. Explanatory Notes from ICE Paper 6662, 'Rate Fixing in Civil Engineering Contracts', by C. K. Haswell, *Proceedings ICE*, February 1963;
3. The ICE Arbitration Procedure (1983).

The Institution of Civil Engineers, the Association of Consulting Engineers and the Federation of Civil Engineering Contractors for the inclusion of the following extracts from the ICE Conditions of Contract:

1. Clause 66 – Settlement of Disputes;
2. Form of Tender and Appendix;
3. Form of Agreement;
4. Form of Bond.

The Association of Consulting Engineers, The Institution of Civil Engineers and the Export Group of Constructional Industries for the inclusion of the following extracts from the Conditions of Contract for Overseas Work mainly of Civil Engineering Construction:

1. Clause 66 – Settlement of Disputes;
2. Notes for Guidance for the Preparation of Conditions of Particular Application, contained at Part II of the Conditions of Contract;
3. Form of Tender and Appendix;
4. Form of Agreement.

The Fédération Internationale des Ingénieurs-Conseils for the inclusion of the following extracts from the Conditions of Contract for Works of Civil Engineering Construction:

1. Clause 67 – Settlement of Disputes;
2. Form of Tender and Appendix;
3. Form of Agreement;
4. Examples of Performance Guarantee and the Surety Bond for Performance;
5. Part II – Clause 34, including example sub-clauses;
6. Part II – Clause 60, including example sub-clauses.

Contents

Chapter 1

The philosophy of the contract
system in civil engineering

The contract system in civil engineering has evolved through the ages for the realization of important capital works for the benefit of mankind. Such works were termed 'civil engineering' as opposed to 'military engineering' in the eighteenth century and have continued to be referred to by this term. The types of work which fall within the ambit of civil engineering are vast: navigable canals, irrigation schemes, roads, railways, docks, harbours, dams, bridges, tunnels and sea defence works may be cited as classic examples. With the advance of technology and the resulting sophistication of the world we live in the scope of civil engineering has expanded. Today this branch of engineering manifests itself in varying degrees in thermal and nuclear power stations, process industries, oil, gas and coal industries, satellite and communication systems and in almost every conceivable project which is established for the benefit of mankind.

The carrying through of a civil engineering project requires that its object should first be identified; then the scheme has to be planned and thereafter implemented. All these stages require not only vision and thought but also organization. The systems used for organization of such a project will depend on its complexity, but its ultimate realization requires the synthesis of three fundamental parties, namely the promoter, the engineer and the contractor. The creation of a contract requires the responsibilities, obligations and duties of these parties to be defined clearly and set out in such a manner that the project is brought into reality properly and with true economy.

At the outset the point must be made, however obvious it may seem, that a civil engineering contract is normally entered into between the promoter and the contractor; the promoter becomes the employer under the definitions expressed in standard conditions of contract adopted for the construction of civil engineering works. Into this context comes the engineer, who provides the technical aspects of designs, specifications and, as we

1

shall see, translates the contract. The engineer's role *vis-à-vis* the employer and the contractor is set out in the construction contract and is normally contained in the conditions of contract that are adopted, remembering that the engineer is not a party to the contract and thus has no legal rights or obligations under the construction contract.

The promoter has as his object what may be termed the seeking of value for money. Policy decisions for the project obviously rest with the promoter. The promoter alone is responsible for the payment of monies due to the contractor as and when they become due according to the terms of the contract: in this context the engineer has to certify the payment within the terms of the contract.

The contractor's object in executing a contract is profit for his shareholders. Within this parameter the good contractor will endeavour to provide the works to a high standard of workmanship and in due time. This enhances his reputation and is obviously a matter of sound policy.

The engineer has essentially two roles to play. The first is his role towards the promoter derived from the contract between him and the promoter under which the appointment was made. The engineer has to ensure for the promoter a finished product that is viable and that the employer is given due value for money. This applies to all stages of the project: the pre-feasibility stage, which usually includes what the Americans term appropriately a 'horseback' estimate of capital cost, alternative schemes, notional times for construction and allied matters; a full and detailed feasibility report; and, during the pre-tender stage, the preparation of tender documents, designs, drawings, etc.; tenders submitted and their assessment; and the detail design and supervision of the construction stage of the contract; maintenance of the works; as-constructed drawings and perhaps operational manuals; and settlement of final accounts and claims. Within the tender period, the period of construction and up to the conclusion of the contract, the engineer has another and equally important role to play as a result of the duties, obligations and functions imposed on him by the construction contract. He has to act impartially and fairly to each of the two parties to the contract, that is, to the employer and the contractor. It also behoves him to be seen to be impartial. This requires skill, tact, experience, wisdom and engineering judgement and, above all, integrity: it is not in any way an easy task for the engineer nor indeed for his representative (or resident engineer as he is usually called) on the site of the works.

The engineer may be a firm of consulting engineers or an individual thereof; or he may be an individual within the organization of the employer. In the latter case the engineer may not readily be permitted to act impartially by his superiors or by the system prevailing in his particular organization; this puts him in an invidious position and may be detrimental to the project. The good employer will be careful to give his engineer full scope to carry out his duties without allowing policy to override the engineer's independence in matters calling for fair decisions.

'The engineer' is an all-embracing term far removed from the legal yardstick, termed an ordinary competent engineer. The engineer in a civil engineering contract is an amalgam of the necessary engineering disciplines, such as structural designers, geotechnical engineers, materials engineers and quantity surveyors – all in the context of technical ability, wisdom, experience and integrity.

Finally it is preferable that the engineer should also have experience of arbitration. Before making an important decision of contract interpretation the engineer will thereby be in a better position to take into consideration how that decision would stand in arbitration.

Turning to the contract itself, it is generally not appreciated that it is a matter of some importance for the correct type to be selected, and this will depend upon a variety of circumstances. Always one seeks to obtain value for money: this is the employer's object. Hence one is also looking for competition among bidders. One needs to provide the all-important incentive to the contractor to provide a good-quality article in due time allied to the opportunity for making a profit.

Again the type of work, the element of specialist design by the contractor, pilot or experimental work and other factors all have a bearing on the best type of contract to adopt.

Unfortunately there is no practical manner by which to decide afterwards whether the right type of contract was adopted: civil engineering is not a repetitive operation and comparisons cannot readily be made. However, experience has shown where, say, a target (*q.v.*) should be applied and where an all-in or package deal contract should not be adopted. There are examples of the latter giving the employer an end product both inferior in quality and greatly over-priced.

The terms of a contract itself are a complex of the rights and duties of the two parties concerned: of knowledge and appreciation of contract risks and contract responsibilities (*q.v.*): of good practice and workmanship within the terms of the contract. These and other major aspects of contract work will be illustrated, as will claims and disputes arising out of contracts. Suffice to say here that one can have no sympathy with spurious claims: cupidity for its own sake is unacceptable. Genuine claims are, of course, a different matter. Apparently ingenuous claims, such as 'the carpenters were not available', must be seen at once as unacceptable. The engineer who is proud to state that he never gets any claim should never be in such a position of responsibility. Certifying additional payment where it is not justified in order to obviate claims shows dishonesty at the expense of the employer. Failure to certify payment when it has been justly incurred by the contractor shows unfair partiality to the employer.

The foregoing is an attempt to paint a general picture of the contract system in civil engineering. As one goes from the general to a particular project it will be seen that there are a variety of technical and financial difficulties to be overcome.

If the tasks of the employer, contractor and engineer are attacked with competence, the outcome is likely to be a successful project. Unfortunately the world is scattered with many projects where the end result falls far short of what may be reasonably expected. It is unfortunately true to state that lack of competence by the engineer or contractor or both is too often the cause.

Chapter 2

Contract promotion

Civil engineering works are promoted in the UK by government departments, local authorities, public corporations, incorporated companies and other bodies such as partnerships and club committees. In overseas countries the bodies that promote civil engineering works are much the same, but often organizations such as the World Bank, the Asian Development Bank, United Nations agencies, development banks and regional organizations play a considerable part in the identification, formulation, financing and implementing of the larger projects in the developing countries.

Capacity of promoters

Government departments

A contract entered into with the British government is a contract with the Crown. The Crown is regarded as a 'corporation sole', which consists of only one member at a time. Contracts with the government are entered through its departmental agents. The Department of Transport, the Ministry of Defence and the Department of Health and Social Security may be cited as examples of agents of the government in the United Kingdom.

Funds for government projects are provided by vote of Parliament; such a vote may be for a particular project or for a series of projects put forward by the ministry or ministries concerned. In overseas countries, especially where the legislature is based on the UK system, often procedures for funding projects similar to that prevailing in the UK are adopted.

The Crown has contractual capacity like any other corporation or subject. In general it is liable to be sued in the same way as a subject. The law governing the contractual capacity of the Crown is contained in the Crown Proceedings Act of 1947. Before this Act the Crown had considerable immunity in the law of contract, which was derived from the common-law maxim that 'the King can do no wrong'. Action for a breach of contract can be brought against the Crown in the High Court or in a county court. The defendant in such a case will be the government department with which the contract is entered into. Where several departments are involved such action will have to be commenced against the Attorney-General, who is the representative of the Crown for purposes of litigation. When entering into contracts with government departments there is no obligation imposed on a contractor to consider whether the department concerned has funds voted by Parliament to finance the contract.

Local authorities

A large proportion of contracts for civil engineering works in respect of public utility services are promoted by local authorities. County, municipal and district councils are typical local authorities. These authorities are constituted under Royal Charter or Acts of Parliament. Contracts are entered into and funded under powers granted in the Charters or Acts constituting them and under general or special Acts governing their procedure and application of funds. Provided a valid contract is entered into, it follows that payments under the contract will be met by the local authority. Contracts cannot be entered into by officials of a local authority unless they are expressly authorized to do so on its behalf. In certain cases a local authority can contract only under a seal. As a rule, a contract of any magnitude with a local authority is sealed with its corporate seal. However, the Corporate Bodies Contract Act of 1960 dispenses with this requirement under certain circumstances.

Public corporations

Promotion of contracts in civil engineering work is effected to a large degree today by public corporations. These corporations are created by Acts of Parliament and are generally for the operation of major industries and services in the public interest. Such bodies, though not part of a government department, are under varying degrees of public control. The corporations are responsible to members of the government such as the Minister of Energy, Minister of the Environment, Minister of Transport and the Minister of Trade and Industry. The Central Electricity Generating Board and twelve area electricity boards in England and Wales; the ten regional water authorities; British Coal; British Rail; and the Post Office are major nationalized industries which fall into this category. The corporations originated in various ways: some as authorities created to discharge functions previously carried out by municipal and privately owned undertakings, as in the supply of gas and electricity; some had the assets of the privately owned coal and railway industries transferred to them; and the Post Office Corporation took on the functions of the Post Office, which grew up as a government department carrying mail which, with the passage of time, had absorbed the privately owned telephone systems. There is a wide variety among the different public corporations and no two corporations are organized in precisely the same way. A public corporation can make any contract which is authorized by the statute of its creation or by any later statutes which have enlarged its powers. Any contract entered into by public corporations outside their powers is void. Again, in a manner similar to that obtaining with local authorities, officials of corporations are not empowered to enter into contracts unless they are expressly authorized to do so on its behalf. Accordingly, contracts of any magnitude with a public corporation bear its corporate seal, although this requirement is no longer necessary under certain circumstances in terms of the Corporate Bodies Contract Act of 1960.

Since 1979, pursuant to government privatization policy, several public corporations have been brought into private ownership. These include British Airways, British Gas, British Telecom, the British Airports Authority and the British Steel Corporation. Among planned privatizations are several water supply authorities and probably some others, including the British Steel Corporation and parts of the electricity supply industry. The statutory monopolies hitherto enjoyed by a number of nationalized industries have now been relaxed by the government.

Registered companies

Much of the contract promotion carried out by the private sector originates from registered companies. These companies are created by incorporation under the Companies Act 1948 or previous and later Acts. Currently the formation and conduct of companies is regulated by the Companies Act 1985, which is a consolidation of Companies Acts 1948–1983. This Act provides for two types of registered company; public and private. The former is not to be confused with a public corporation created by Act of Parliament. A registered company is fully liable for its debts but the liability is generally limited. The limitation may be to the amount unpaid on the members' shares as obtaining in a company limited by shares or to the amount they have agreed to pay if the company is wound up, as is the case with a company limited by guarantee. Some companies, although a minority, are unlimited, and in such cases their members are fully liable for unpaid debts of the company. The capital of a public company is usually raised by the issue of shares to the public. When making such an issue it is a statutory obligation to issue a prospectus detailing full particulars of the company including, *inter alia*, capital structures, profit records, names of directors and many other particulars so as to assist prospective shareholders to assess the company. The powers of a registered company are contained in its Memorandum of Association and the Articles of Association, both of which are lodged with the Registrar of Companies at the time of incorporation. The Memorandum governs the relations between the company and the outside world and includes the important 'objects clause', which sets out the activities in which the company can legally engage. The Articles of Association contain the regulations that govern the relationship between its members and the company and as such set out the internal affairs of the company. Both the Memorandum and the Articles of Association are open for inspection by the public. The company must act within the Memorandum and the Articles. The objects clause in the Memorandum which sets out the company's contractual capacity can be altered after incorporation to add to or diminish this power. The European Communities Act 1972 deems any transaction decided upon by the directors of a registered company as being within the powers of the company. It follows that, provided a contract is entered into by the directors of a registered company, the *ultra vires* doctrine would not apply. The 1972 Act does not affect other corporations.

The major undertakings in the private sector are public registered companies. British Petroleum, Imperial Chemical Industries, the General Electric Company, Blue Circle Industries, the Rugby Group and the Eurotunnel Group may be cited as typical major private sector companies.

Other bodies

A body of individuals, club committees and partnerships are unincorporated associations. These may conduct their affairs in much the same way as incorporated associations but they do not possess an independent legal personality. Thus an individual who makes a contract on behalf of such an association is usually personally liable. It follows that when entering into a contract with such bodies the authority of the individual or individuals to bind the others in personal liability must be examined. Before entering into a contract with such bodies or an individual suitable references should be obtained to ensure that their individual financial status is adequate for effecting payment under such a contract.

Funding of projects by the promoter

It is axiomatic that funds required for a project have to be provided by the promoter.

Contracts promoted by government departments are mainly in respect of projects concerning the national infrastructure. Defence installations, roads, bridges as well as office complexes connected with government administration are some examples of such projects. The funds for these projects are derived by the government from monies raised through the public finance system by direct taxation of companies and individuals, indirect taxation such as excise duties, value added taxes and stamp duties, public sector borrowing and other means such as Treasury Bills, Government Securities and various forms of national savings. The government surveys the whole range of public expenditure each year. The survey takes the form of the expenditure forecast for the budget year and four years ahead with comparative figures given for a corresponding number of past years. A report on the survey is prepared by Treasury officials and other government departments under the direction of the Public Expenditure Survey Committee, which is presented to the ministers. When ministers have reached their decisions on the public expenditure plans they are published as a White Paper and provide a basis for the public expenditure debate in Parliament.

The department responsible for the project prepares estimates for them, sometimes employing outside consultants. These estimates are included in the department's estimates of cash requirements and submitted to the Treasury in the December before the financial year beginning on the following 1 April. After Treasury approval these estimates, which, where necessary, are brought before the Cabinet, are embodied in the government's supply estimates which are presented to Parliament in February or March shortly before Budget day. These are debated on broad policy issues rather than detail during the 29 days allotted in each session known as the supply days. The supply estimates are approved by Parliament for the financial year ahead by means of the annual Appropriation Act in July. The expenditure to this date is authorized by a vote on account approved by the government at the beginning of the financial year. There are certain expenditures which are not voted annually but are covered by separate Acts of Parliament allowing payments to continue from year to year from monies drawn on the Consolidated Fund.

Expenditure by government departments is monitored by the departments concerned and the Treasury. These are also subject to scrutiny by the Auditor-General, the House of Commons Select Committee on Expenditure and the Public Accounts Committee, which is also a House of Commons Select Committee whose chairman by tradition is a member of the Opposition with ministerial experience where this is possible.

Contracts promoted by the local authorities are generally in respect of public utility services such as construction of sewers, roads, buildings and similar services provided mainly for the benefit of residents within the jurisdiction of the authority. Projects such as main roads, tunnels and sea defence works which are for the benefit of the population at large are also undertaken by the local authorities often as agents of the government. The funds for local authority projects are obtained from general rate levies, supplementary rate levies, direct grants by the government and from loans obtained by the government and other institutions: these are subject to change when the proposals for the reform of local government finance announced in 1986 are eventually implemented. The direct grants required from the government are included in estimates of cash requirements presented to Parliament annually through the Treasury; these, however, are not recorded as public expenditure until the money is actually spent. The local authority borrowing met by the government is dealt with through the Public Works Loan Board, which has recourse to the National Loans Fund. The local authorities also borrow directly from the market either on a short- or a long-term basis through a variety of investments.

A variety of civil engineering projects is promoted by public corporations. On the one hand, organizations such as water authorities, the Housing Corporation and new town development corporations concern themselves with capital projects their names signify: on the other, the nationalized industries such as the Central Electricity Generating Board, British Rail or a water authority would promote projects such as a new power station, a rail bridge or a pumped storage scheme. The Acts setting up public corporations define their financial obligations. Once the initial capital is met by the government the corporations are generally expected to carry out their operations on a commercial basis. This would mean that their revenues should at least cover all items properly chargeable to the operation, including interest, depreciation, the provision of reserves and, in some cases, redemption of capital. The funding of contracts promoted by public corporations can be from their own reserves, government grants and loans or borrowing from the market. The bulk of borrowing by public corporations is met by the central government through the National Loans Fund. Temporary borrowing by public corporations is met largely from the market backed by a Treasury guarantee.

Public corporations, being essentially autonomous commercial organizations, are subject to government financial control upon the degree to which they are dependent on public funds. The accounts of public corporations are audited by professional auditors appointed by or with the approval of the appropriate minister. Public corporations' accounts are not audited by the Auditor-General, as is the case in government departments, but he can be present to draw attention to any point which occurs to him when such accounts are examined by the House of Commons Select Committee on Public Accounts. Public corporations' accounts are presented to Parliament as a part of the annual report or otherwise. Unlike government departments, public corporations do not have to submit an estimate to Parliament nor do they have to observe the complicated formalities to which government accounts are subjected through the interaction of the Treasury, the Comptroller and Auditor-General and the Public Accounts Committee. Although there is public accountability the public corporations are in a position to evolve their accounting and control system in a flexible manner to discharge their functions efficiently.

Registered companies are responsible for promoting a very wide range of civil engineering projects. At one end of the scale these could be large oil refineries, oil platforms or a huge industrial complex and at the other a small cycle-shed or a boundary wall around a factory. The funding of the capital projects is generally met from the reserves of the company concerned which are augmented as appropriate by borrowing from the market. Capital is raised by newly formed registered companies by public subscription for its shares. All registered companies are required under the Companies Act to file an annual return with the Registrar of Companies giving, *inter alia*, particulars of share capital, debentures, mortgages and charges, together with a statement of the company's accounts and the auditors' report. The auditors are appointed by the shareholders at the annual general meeting of the company.

The nature and degree of internal financial control adopted by registered companies are determined by its board of directors. The scale and nature of the company dictate the strength of the financial management and control systems that are adopted by the individual organizations.

In the recent past in Britain and overseas, private initiatives for what were more usually public sector undertakings have been responsible for promoting and funding large engineering projects. The basic principle governing the establishment of such projects is that a concession is granted by the government to a prospective company to operate and take profits for a specified period on a project which has to be financed, designed and built by that company. The assets of the company revert back to the government after the specified period. The companies which establish such projects are normally consortia comprising construction companies, merchant banks, commercial banks and others who have a particular expertise and the capacity to contribute towards the undertaking. Examples of such projects in Britain are the Eurotunnel and the second Dartford River Crossing.

The scale of individual projects promoted by unincorporated bodies such as partnerships, club committees and by individuals is small. However, the number of such projects can be said to be very large. The funds for these projects are raised by subscription, donations and, in the case of an individual, from his savings or loans from a building society or a bank. There is obviously no standard set of rules laid down for the control of funds generated by these bodies. It follows that the manner of funding in each case should be ascertained and a contractor normally gets paid at the right times. The risk is somewhat reduced when such work is channelled through an independent professional adviser such as a consulting engineer, an architect or a chartered surveyor.

International financial institutions

Promotion of large civil engineering projects, especially in the developing countries, is dependent to a large degree on loans granted by international financial institutions. The World Bank was for some years the only institution of this kind, but during the last three decades several regional institutions have been established progressively to discharge this function. These include the Asian Development Bank, the Inter-American Development Bank (for development in Latin America), the Arab funds, the African Development Bank, the OPEC Fund for International Development and the European Development Fund (for development aid under the Lomé convention and the Mediterranean Agreements).

The policies, rules and procedures of these institutions differ from one to another, although the main purpose for which they are all established is to lend funds for productive projects in a developing country, generally to foster economic growth in the region for the benefit of all. A brief description on the manner in which the World Bank and the Asian Development Bank were established and are operated is set out below.

The World Bank

The World Bank was established in 1945 following the Articles of Agreement or Charters, which were drawn up by the United Nations Monetary and Financial Conference held in July 1944 at Bretton Woods, New Hampshire, USA. The Articles of Agreement drawn up were for two complementary financial institutions, namely the International Monetary Fund (IMF) and the International Bank for Reconstruction and Development (World Bank).

The roles assigned to the two institutions differed, although their common objective was to provide the monetary and financial machinery that are required for nations to work together towards world prosperity, thus aiding political stability and fostering peace among nations. The Bank's main concern is with economic development whereas the IMF deals with monetary affairs.

The UN Monetary and Financial Conference was convened by the 44 allied nations during the Second World War, and of these governments all except the Soviet Union became members of both the World Bank and the IMF. The Bank is now owned by the governments of more than 150 countries. The capital of the Bank is subscribed by its member countries and finances its operations from its own borrowings in the world's capital markets, with a substantial contribution to its resources coming from its retained earnings and flow of repayments of its loans.

The World Bank in its present form is a group of three institutions, namely the International Bank for Reconstruction and Development (IBRD, established in 1945) and its two financial affiliates, the International Development Association (IDA, established in 1960) and the International Finance

Corporation (IFC, established in 1956). The only non-financial affiliate of the Bank is the International Centre for Settlement of Investment Disputes (ICSID), which provides a voluntary mechanism for settling disputes between governments and foreign investors. The headquarters of the World Bank are in Washington, DC, USA.

The IBRD's first loan made in 1947 was for the post-war reconstruction in four European countries shattered by the war. With the advent of the US Marshall Plan in 1948 the Bank turned its efforts mainly to development lending. The Bank's main function today is lending for productive projects which will lead to economic growth in less-developed countries. The IBRD grants loans normally for specific projects which are technically and economically sound and of high priority for the economic development of the country. The loans are generally repayable over 12 to 15 years or less and have a grace period of three to five years. The interest rate charged on loans is calculated in accordance with a formula relating to the Bank's cost of borrowing. This rate has been consistently lower than that countries would have to pay if they were to borrow directly from the private capital market. To obtain a loan from the Bank the project should be in a member country or in a territory under the administration of a member. A prerequisite to the grant of a loan is that the project is well managed both during its implementation and after completion, and that there is a reasonable assurance that the loan will be repaid and does not impose an undue burden on the borrower

The International Development Association (IDA) is the affiliate through which concessionary loans (called 'credits' to distinguish them from Bank loans granted on conventional terms) are given to the poorest countries for high-priority economic development projects which are sound and productive. These credits are granted without interest or at a very low rate of interest and repayable over very long periods. The Articles of Agreement provide that the IDA terms should be 'more flexible and bear less heavily on the balance of payments than those of conventional loans'. IDA credits to date have been for a term of 50 years and bear no interest. The period of grace is ten years, after which 1 per cent of the credit is to be repaid annually for ten years while in the remaining 30 years 3 per cent is to be repaid annually. There is an annual service charge of 0.75 per cent on the disbursed portion of the credit to cover administrative costs. IDA credits are only made to governments and they are directed mainly to those with a per capita gross national product of less than US $790 (in 1983 dollars). IDA's funds are obtained from five sources; members' initial subscriptions; periodic 'replenishments' provided by its richer members; special contributions made by some members; transfers of income from the IBRD; and the IDA's own accumulated income. Although legally and financially the IDA is distinct from the IBRD it is administered by the same staff.

The International Finance Corporation (IFC) is the affiliate of the Bank which concerns itself with promoting the growth of the private sector economies of the less-developed countries. The stated purpose of the IFC is to encourage, stimulate and support the growth of productive private enterprise and domestic capital markets in developing countries. The IFC invests with other investors, local and foreign, in projects that establish new business or require expansion, modernization or diversification of existing business. It considers investments from the point of view of an investment banker and that of a development institution. The IFC, apart from granting loans, provides assistance to private organizations on technological and management aspects, makes equity investments and provides standby or underwriting commitments. Its funds are derived from the subscribed capital, borrowing from the World Bank and other sources, and from repayment of loans and the sale of its investments. Sales of its investments also help to stimulate capital markets in the developing countries and contribute to the international flow of productive investment. The IFC has its own operating and legal staffs, except for the President, and its own management under the direction of an Executive Vice-President. Its operations are closely connected with those of the Bank from which administrative and other services are also drawn.

The International Centre for Settlement of Investment Disputes (ICSID) is the only non-financial affiliate of the Bank. This institution provides a voluntary mechanism for settling disputes between governments and foreign investors. The jurisdiction of the ICSID is based on consent given by both parties to an investment dispute, i.e. between the private foreign investor and the relevant member government.

The World Bank, which was established before the United Nations, became a specialized agency of this body in 1947. The agreement recognizes the operational independence of the Bank from political bodies of the UN. A close relationship exists between the World Bank and the UN. The President of the Bank is a member of the Administrative Committee on Coordination, the chairman of which is the Secretary-General of the UN and whose other members are the heads of other specialized agencies and major UN 'Programmes'. The Bank has constitutional links and special responsibilities in relation to nearly all other specialized agencies and major programmes of the UN system.

The Asian Development Bank

The Asian Development Bank (ADB) was established in 1966 following a resolution endorsing a proposal to establish a regional development bank for Asia at the first Ministerial Conference on Asian Economic Cooperation. This was held in Manila in December 1963 under the auspices of the Economic and Social Commissions for Asia and the Pacific (ESCAP), then known as the Economic Commission for Asia and the Far East (ECAFE). The Bank's headquarters are in Manila, in the Philippines.

The Bank's stated purpose is to foster economic growth and cooperation in the regions of Asia and the Far East, including the South Pacific. The Bank is owned by its member countries of which 31 are from the region and 14 from outside. The 31 regional members comprise 28 developing countries, both large and small, and the remainder are developed countries from the region; and 14 industrially advanced and major capital exporting countries from Western Europe and North America. Sixty-two per cent of the capital of the Bank is held by the regional members with the non-regional members holding the balance; the Bank's Charter provides that a minimum of 60 per cent of the capital stock be held by members within the region. Membership of the Bank is open to members and associate members of the ESCAP and to other regional and non-regional developed countries which are members of the UN or any of its specialized agencies.

Like the World Bank, the ADB makes two kinds of loans: ordinary loans to the somewhat better-off countries and concessional ones from its special funds to the poorest member countries; sometimes loans constitute a blend from its special funds and ordinary capital resources.

During 1983 the Bank's lending rate from its ordinary capital resources was 11 per cent per annum; the maturity period varied from 10 to 30 years with a grace period of two to seven years. The concessional loans carry only a service charge of 1 per cent, with repayments extending over 40 years, including a grace period of 10 years. The Bank's ordinary loans are met from subscribed capital, funds raised through borrowings and its reserves. Concessional loans from its special funds are met from up to 10 per cent of the paid-in capital of the Bank authorized under its Charter, voluntary contributions made by members, income on special fund loans and income earned by undisbursed special fund resources. The Bank currently administers two special funds: the Asian Development Fund (ADF) established in 1974; and the Technical Assistance Fund (TASD). The special funds of the Bank are managed separately from the ordinary capital resources. With a progressive increase in concessional lending additional funds have been mobilized on an organized and regular basis. The Bank's funds for concessional lending are administered through the ADF.

The Bank, apart from project financing, lays considerable emphasis on providing technical assistance to developing member countries to improve their capability to make use of external project financing, whether this is provided by the Bank or by other sources. Its technical assistance activities cover project preparation, project planning, project implementation, sectoral studies, development planning and other economic aspects of national or regional concern.

The ADB is not a member of the United Nations family of organizations. However, the Bank maintains a very close relationship with the UN and its various organs and specialized agencies. The Economic and Social Commission for Asia and the Pacific (ESCAP) and the Secretary-General of the United Nations played a leading role in the establishment of the Bank, and the cost of the preparatory work connected therewith was funded by the UN Development Programme (UNDP), which is a specialized agency of the UN.

Project identification and formulation

Project identification

The very first step towards the establishment of a project is to identify its need. Projects have been identified by men to attain their objectives, whatever they may be, since the advent of civilization. But as human society developed and mankind's needs and wants grew, the establishment of projects to satisfy these aspirations became even more essential and, by necessity, more complex.

Today with the advances in business and industrial technology, the wide growth of markets, products and services and the need for the full development of the resources of all nations, more and more projects are identified as being necessary to enhance the living standards of people throughout the world.

The need for a project can be identified by anyone. When a man found that he could not cultivate the field adjoining his because the two fields were separated by a stream the need for bridging the stream had been identified. Likewise, the need for the construction of an oil platform, a long-span bridge, a main highway or an irrigation scheme is identified by the party or parties that wishes this need to be fulfilled. The decision as to whether resources should be employed for this purpose lies with politicians, administrators, industrialists and others who often in the present-day context bring to bear the expertise that is available from economists, engineers, financiers, bankers and others even before a detailed study is commissioned on the ways and means that are available for the establishment of the project.

Project formulation

Having identified a project, the next important stage, during which an in-depth study is carried out on whether the project should be implemented, is often defined as the *project-formulation* stage. During this period a *preliminary study* is first made of the alternatives that may be available to realize the project that has been identified. This is followed by a *project-feasibility study* upon which the decision will be made as to whether the project is to be implemented or not.

The form of the preliminary study undertaken varies in relation to the nature of the project. The object of this study is to find out whether the project is economically justified before embarking on the more costly feasibility study. It is not always the case that lack of an economic justification thwarts the realization of a project; a social benefit which the project may engender could necessitate its realization. At all events, it is necessary for the party who wishes the project to be established to have a rough idea of its viability derived from all the basic data, however approximate they may be, before a decision can be made to move on to the project-feasibility study stage.

The matters that have to be taken into account at the preliminary study stage for the establishment of a manufacturing industry would have to cover a market survey, the technology processes involved, availability of raw materials, power, manpower and other relevant data necessary for the manufacture and marketing of the product envisaged.

After having collated the necessary preliminary data the likely capital cost for locating the works at each of the alternative sites is evaluated. These are done on a very broad basis of elemental costs for providing the facilities. However, these costs have to be worked out realistically and should take into account all factors such as costs of services of specialist staff, materials and plant that have to be procured both locally and from foreign sources. It is important that the costs should be valid at the date on which the project study is prepared. An indication of the increase of these costs on the probable date for the commencement of the project, and during its construction period, owing to inflation, price fluctuations in materials and labour and the like should also be forecast for the benefit of the promoter.

In addition to working out the capital costs an estimate of the operational cost, production and sales forecasts, expected revenue, the working capital required and other financial data should be included in the preliminary study. There are no particular rules which set out the degree of precision to which all these elements should be examined, but the preliminary study has to be sufficiently complete and realistic to enable the promoter to make a decision as to whether the next stage required for formulating the project is to be undertaken or not.

The next stage of project formulation is the preparation of the *project-feasibility study*. This study has to set out and analyse all aspects of the project in sufficient detail to enable the promoter to make a decision as to whether the project should be implemented or not. Depending upon the scale, complexity and type of project, the matters that are considered in a project-feasibility study vary. The promoter's object in establishing a multipurpose irrigation scheme will not be the same as would be the case for the establishment of a manufacturing industry, a housing and shopping development or a public transport system. On the one hand, the return on investment may not be required because the project has a social benefit, but, on the other, the sole object would be a quick return on the investment so that the expenditure incurred on the project could be recovered as soon as possible. The techniques available for project evaluation and appraisal are numerous. Often different methods may be adopted for the same type of project, but if the types of project are not similar, by necessity, different methods are likely to be used. However, the basic tenet underlying the preparation of a project-feasibility study could be stated to be common for all projects. Accordingly, some aspects concerned with such a study as applicable to the manufacturing industry are outlined in this chapter.

The matters that are covered in a project-feasibility study for a manufacturing industry such as a cement-making plant in a developing country may be expressed in the following terms:

1. Detailed market analysis, including pricing levels;
2. Raw material analysis;
3. Engineering study to cover process, cement manufacturing plant, mechanical and electrical installation and civil engineering works;
4. Detailed analysis of manpower requirements;
5. Detailed analysis of off-site utilities;
6. Estimate of capital costs;
7. Estimate of projected sales revenue, production costs and anticipated working capital requirements;
8. Profitability analysis;
9. Commercial profitability;
10. National economic profitability where applicable;
11. Programme for the establishment of the project.

The above consists mainly of an elaboration of matters that had been covered in the preliminary study.

The market analysis for a cement works, for example, would involve research on the present consumption and a forecast of the future demand patterns. The former is relatively straightforward. The quantity of cement produced locally can be ascertained as a first step. The quantities imported into the country annually can be obtained from a governmental organization such as the import control department or other sources that may be available. The areas in which the cement is used will also have to be ascertained. Details of the retail distribution networks for the product as well as volumes of sales at these outlets and the purposes for which the cement is put to use will have to be elicited. Bulk sales made to contractors for major civil engineering and building works will have to be recorded. The compilation of data on future demand patterns is more onerous and carries an element of risk. The future demand of cement depends to a large extent on the economic growth of the country concerned: the growth indicators of the country, its Gross National Product, the Net National Product, per capita incomes and consumption and such factors will have to be studied. A judgement on objective criteria will have to be made of civil engineering and building works envisaged for the future. The elements of the pricing structure of cement obtaining in the country such as ex-works prices, CIF prices of imported cement, customs duties imposed, transport costs, sellers' margins and other relevant components should be ascertained and assessed. The market analysis should be carried out by a team well versed in economics, statistics, local knowledge and, above all, commonsense, so that the analysis is accurate in respect of present consumption figures and realistic in relation to future demand patterns.

The raw material analysis undertaken at this stage should be detailed. Not only should the survey include the volumes of raw materials that are available for production of cement but it should also cover the methods to be adopted for its extraction, storage and transport, as well as the physical and chemical properties of the material to enable the optimum process required for its conversion to cement to be determined.

The engineering study contained in the project-feasibility report should be sound. The best possible expertise that is available locally or internationally should be harnessed for the preparation of this fundamental element of the project study. If the project under study is a new development within an existing cement-manufacturing organization much of the expertise that is required for this purpose may be available from within. The choice of the correct manufacturing process in relation to the available raw materials, fuel, water and manpower resources is of paramount importance. The manufacturing process should be the correct one, which will not necessarily be the cheapest to buy; the operational costs of the various processes have to be evaluated as well as the availability of spares and the expertise that would be available in the country concerned for repairs and maintenance of the plant and equipment when they are installed. The study should also cover elements such as mechanical and electrical equipment, civil engineering works, ancillary and auxiliary buildings and equipment, such as the requirements of office buildings, laboratories, stores, mechanical handling equipment, haulage and raw material extraction equipment, laboratory equipment, initial spares and all such facilities as are required for the production of cement.

The analysis of manpower requirements at this stage is vital, as on this factor the efficient operation of the works mainly depends. The organization structure from top management downwards should be carefully worked out. The cadres of the various divisions such as production, maintenance, accounting, personnel, supplies, research and development, marketing, sales and security have to be carefully prepared. When the requirements are known it is necessary to assess the availability of persons in the market to fill the posts. In many instances extensive training may be required before the manpower requirements of the plant can be satisfied. These aspects should be brought out clearly in the manpower analysis with full cost forecasts for the manpower requirements necessary for the operation of the plant.

The availability or otherwise of off- and on-site utilities at a particular location has a considerable bearing on siting a factory. For instance, if water supplies, power, rail heads and developed transportation systems both for inward supplies of raw materials and for despatch of the finished product are readily available these facilities do not have to be developed. Perhaps the charges for these services will have to be met and taken into consideration in computing operating expenses, but the capital cost and the time required for setting up these facilities will be saved. Often in developing countries these facilities are not available, and consequently the project study should examine and assess these aspects. The capital cost and the time required for establishing these utilities can be considerable; sometimes the cost of providing these may be as much as 40 per cent of the cost of the project. In planning the utilities the adequacy of these facilities for the future expansion of the works should be considered. For instance, in the longer term the establishment of a large power station or a water supply scheme to meet the immediate and projected requirements could prove more economical.

The estimate of capital costs is prepared following a thorough examination and assessment of the engineering aspects of the project. Frequently more than one scheme is considered at the project-feasibility study stage. It follows that if options are considered capital costs in respect of each will have to be given. A typical list of items against which costs are placed in a capital works estimate for a cement manufacturing plant would be as follows:

1.   Land and site development
    1.1 Land acquisition
    1.2 Site clearance and levelling
    1.3 Boundary fencing
2.   New cement-making plant (cost of plant f.o.b.)
    2.1 Mechanical plant
    2.2 Electrical plant
    2.3 Initial spare parts
3.   3.1 Freight
    3.2 Insurance
4.   Unloading and transportation of equipment to site
5.   Plant erection and commissioning
6.   Civil engineering construction complementary to plant
7.   Plant auxiliaries
    7.1 Workshop building
    7.2 Workshop equipment
    7.3 Internal mechanical and handling equipment

8. Auxiliary buildings and equipment
   8.1 Administration block
   8.2 Laboratory
   8.3 Laboratory equipment
   8.4 Stores building
   8.5 Miscellaneous buildings
   8.6 Weighbridges
   8.7 Motor lorries and vans
   8.8 Passenger cars and other vehicles
9. Quarrying, raw material handling and transport equipment
   9.1 Drilling equipment
   9.2 Quarrying equipment
   9.3 Raw material transport equipment
   9.4 Shunting locomotives and wagons
10. Factory services
   10.1 Railway sidings
   10.2 Internal roads
   10.3 Sewerage and sewage disposal schemes
   10.4 Factory-lighting and distribution services
11. Off-site utilities
   11.1 Water supply scheme
   11.2 Power station
   11.3 Sub-stations and electrical feeder network
12. Welfare and housing
   12.1 Canteens, welfare blocks and changing rooms
   12.2 Medical centre
   12.3 Club houses, sports and recreation facilities
   12.4 Housing schemes for employees
13. Engineering and design fees
14. Staff training expenses
15. Contingencies.

The total cost for the above items would represent the capital cost of the project. These are broken down into *foreign costs and local currency* in respect of projects located in countries where the local currency is not freely convertible or where currency restrictions apply. The proportion of the capital cost that is envisaged to be expended during each year of implementation should be set out; this will have to accord with the overall programme prepared for the implementation of the project.

The estimate of *sales revenue* is an estimate of the returns from the sale of goods produced at the works, and is mainly based on the *market analysis* prepared for the project to assess the likely production. Concurrent with the estimate of sales revenue, the cost of production of the saleable goods has to be worked out. In computing the production costs they are broken down into variable costs and fixed costs; variable costs are those which are proportional to the volume of production, whereas fixed costs are those which would be incurred whether or not goods are produced. This information is collated and analysed by a study team comprising production engineers, cost accountants, marketing management and financial consultants, after which a profit forecast is prepared for the project for each year of commercial production until production reaches its desired level. As is the case with all elements of the project-feasibility study, the profit estimate should be worked out realistically. However, with the establishment of a new enterprise the estimated revenue can vary with market fluctuation and consequently the production capacity may have to be adjusted; there could also be variations in anticipated production due to many causes. In order to view the profit under such eventualities the *break-even* point for the operation is worked out. The break-even analysis is a useful device by means of which the effects upon costs and profits may be predicted due to variations in production or sales volume. The manner in which the break-even point can be determined graphically is shown in Figure 2.1.

In Figure 2.1 the sales revenue is shown by line AB, fixed costs by AC and variable costs by CD. The break-even point in this example is X, where 3000 units produced are sold. At this point the sales revenue equals production costs, the value of both being £500 000. The difference in ordinates of the triangle BXD represents profit and those within triangle ACX show the magnitude of loss for the different levels of units produced and sold.

The *working capital* requirements in a manufacturing industry may be stated to be the monies tied up in the enterprise for materials and consumables required for production and for finished products which have not been sold. These comprise raw materials, other materials required for conversion of raw materials to the finished product, fuel, packing materials, spare parts, work in progress and products not sold or for which payment is yet to be received. The working capital requirement has to be assessed on the optimum levels of these items that have to be available at any one time at the works. Provision of working capital has to be taken into account in the project-feasibility study.

The method by which the overall profitability of a project or that of competing projects are evaluated is often termed the *profitability analysis*. The techniques used for this analysis are

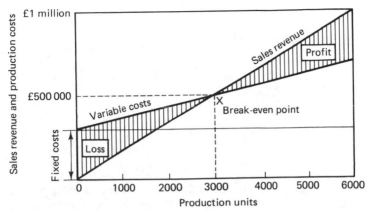

Figure 2.1. Break-even analysis

many and are of varying degrees of sophistication. Some of the techniques available are the *pay-back* and the *discounted cash flow methods*. The pay-back method is, in essence, the determination of the number of years taken for the total investment to be paid back in the way of profits. The discounting methods generally fall into two categories: *net present value* and the *internal rate of return*. These take account of the costs and benefits of a project throughout its life by discounting these to the present date.

The commercial profitability of an enterprise is also worked out whether the project may be one within the private sector or one falling within the public sector. However, it is often difficult to place money values on all costs and benefits of a project to be established by the public sector for the benefit of the community at large. Today economists have devised ways and means of assessing the monetary values of the costs and benefits that projects engender by means of cost-benefit analyses. In many cases this has been possible, but in projects such as defence installations and the like this exercise has to be one of political judgement alone.

The foregoing illustrates some of the matters that are examined and analysed for the compilation of the project-feasibility study. The promoter bases his decision as to the implementation of the project on this study. When this fundamental decision is made the next step that the promoter has to take is to find the finance that is required for the project. After that is secured he embarks on the implementation of the project for which normally one or more contracts are let. The chapters that follow discuss and illustrate the many aspects involved in this task, particularly as applicable to civil engineering contracts.

Chapter 3
Types of contract

A *contract* is essentially an agreement between two or more parties to do or to refrain from doing something. For a contract to be legally enforceable under English law there should be an intention to create legal relations. A contract stems from an *offer* and *acceptance* and the intention of the parties to create legal relations is required to be evidential by law in the way of *form* and *consideration*; the latter, however, is not required if the contract is made under seal or in Scotland. When these requirements are satisfied a contract is formed. For such a contract to be valid and enforceable the other essential elements which have to be satisfied may be expressed in the following terms:

1. The capacity of the parties to contract.
2. The genuine nature of consent of the parties who make the offer and those who accept it.
3. The objects of the contract should be legal and possible.

Civil engineering contracts are invariably formed by a process of offer and acceptance, both of which are in writing. The offer is made by a contractor to an employer to carry out the work set out in contract documents normally on the form of tender issued to him. A simple contract is formed when a tender submitted by a tenderer (which is the offer) is accepted by the employer. A point to note is that for a contract so to be formed the tender and the acceptance should be unconditional. Often when tenders are submitted several qualifications are attached to them; in other words, the offer is not made strictly in accordance with the contract documents. In such cases it is usual for the qualifications to be discussed and clarified between the contractor and the engineer (acting on behalf of the employer). After such clarification either the tenderer is asked to withdraw such qualifications or these are embodied in notes as agreed statements which constitute an integral part of the offer: it is essential that the original offer is kept alive during discussions after tenders are received, as a counter-offer extinguishes the original. When accepted by the employer such a revised offer will then form a simple contract.

In civil engineering contracts provision is normally made for an agreement under seal to be executed if the employer elects to do so. When such an agreement is executed the simple contract merges into a deed.

For a contract executed under seal the limitation period is 12 years, whereas for a simple contract it is six years. The limitation period is the time within which an action for breach of contract must be commenced, and this period starts to run when the breach occurs and not from the date on which the contract is made. The right of action after the expiration of these periods is barred under the Limitation Act 1980.

Civil engineering work is usually executed under a contract entered into between the employer and a contractor.

Contracts may be classified as follows:

1. Admeasurement contracts, including
   (i) bill of quantities, or
   (ii) schedule of rates.
2. Lump-sum contracts.
3. Cost/reimbursement contracts
   (i) cost plus percentage fee, or
   (ii) cost plus fixed fee, or
   (iii) cost plus fluctuating fee
   (Application of a target may be made to the above when the contract is termed a 'target contract'.)
4. Design and construct contracts.
5. All-in contracts (also called 'package' or 'turnkey' contracts). The all-in contract is normally a lump-sum contract.
6. Management contract.

As an alternative to execution of work by contract the employer may resort to the method known as *direct labour*. Under this system the employer uses labour and deployed plant already available to him or obtained specifically for the project. He undertakes all the responsibilities of management and risks of construction and pays all wages costs and expenses as they are incurred. There is no contract as such although the employer will be required to discharge all statutory obligations in carrying out the construction work.

Admeasurement contracts

Bill of quantities

The usual form of admeasurement contract embodies a bill of quantities where detailed lists of all items of work required to be carried out with approximate quantities are prepared, against which the tenderer enters unit rates or prices. The tender total is the aggregate amount of various quantities priced at the quoted rates or prices, together with such extraneous items as provisional sums and so on. As the work proceeds the actual quantity executed under each item is subject to admeasurement at the quoted rate. The items of work are normally listed under separate headings for which a bill number is given. It is quite usual to provide contingency sums and provisional sums for works not fully identified in the bills. It is also normal to provide separate bills of preliminaries, general items and dayworks in this form of contract.

This type of contract is classed as admeasurement because the unit rates tendered by the contractor for individual items are fixed, although it does not give a fixed total sum as the quantities are measured and extended on an actual basis. Provision is made for valuation and adjustment of rates and quantities for varied, additional work or extra work.

Some of the advantages attributed to this method are:

1. The contractor is paid for the amount of actual work he does.
2. While remaining a fair basis for payment there is freedom for the alteration of work.
3. Adjudication of tenders is relatively easy in that all tenders are priced on a comparable basis.
4. The tenderer is given a clear conception of the work involved by the way of bills.
5. Most contractors in the UK at least are familiar with this type of contract and consequently are in a position to price the work in a fair and reasonable manner.

Schedule of rates

Where a schedule of rates is adopted it is usual for this to be in effect a comprehensive list of the various items of work to be carried out. No quantities are given and, indeed, the only quantities which are relevant are final quantities. It may be that the contractor is invited to propose his rates for the items or to quote a percentage above or below rates used on a previous contract or, indeed, rates previously entered into the schedule by the engineer. Except in special cases where full data are available to the contractor the use of a schedule of rates is not ideal, since it is difficult for the contractor to arrive at a realistic on-cost at a time when the quantities are unknown. However, it is usefully employed on special types of work such as maintenance projects. It is not suggested that in any major project the adoption of a schedule of rates is in any way comparable with the adoption of a bill of quantities.

Lump-sum contracts

A lump-sum contract has been used for many years and is especially popular in the USA. Indeed, the tradition of the civil engineer in America has been influenced by the necessity of the engineer to finalize all the details of the work to be undertaken at the tender stage. This is a highly commendable discipline but, unfortunately, it has not extended to the UK. The opposing theory is that the later one waits until one completes a design, the more advantage one may take of the advances of the design in such matters as plant. However, it appears that the latter has been taken too far in the UK, and an example of this may be taken if we consider the modern thermal generating station. In the UK the time for construction is generally five years whereas in the USA it is of the order of four years. The reason for this is quite simply that no one has been able to force electrical and mechanical engineers in the UK to finalize their designs at the right and proper time. At all events, much work in the USA is carried out under the lump-sum contract, and one of the main prerequisites is that one must know exactly what one is to construct at the time of going to tender. It behoves the employer thereafter not to change his mind, since alterations on a lump-sum contract are expensive.

It should be pointed out that this type of contract is popular with financial advisers to employers on the premise that one obviates extra expenditure in relation to the admeasurement contract. This ignores the question of changes during the contract period mentioned above and often involves inflated contract prices. Again, claims are not necessarily excluded.

In all other respects a lump-sum contract follows much the same lines as an admeasurement type of contract.

Cost/reimbursement contracts

The cost/reimbursement contract has been known in the past as a cost-plus contract and in earlier days by the much more elegant form 'time and lime contract'. It is, of course, analogous to the works carried out by dayworks, which forms a discrete and ancillary part of the admeasurement contract.

This type of contract came into its own during the First World War and, regrettably, from that time onwards it bore something of a stigma based on the hypothesis that this was merely an excellent way for contractors to make money.

It was used to a large extent during the Second World War, but endeavours by this time had been made to try to improve both the image and effectiveness of such a type of contract. This was achieved by applying a target, although it must be admitted that even with the best of targets it is difficult to apply the essential ingredient in any civil engineering contract, which is that there shall be an incentive for the contractor to carry the work out both expeditiously and correctly.

The basis is the reimbursement to the contractor in the various categories, such as salaries and wages for his employees; plant by means of a hire schedule; materials both permanent and temporary by the settlement of the various accounts which such materials draw for themselves and usually, say, 2½ per cent, for small tools; and the actual cost of such extraneous matters as fuel and power. The total of the foregoing may have either as a percentage or a fixed fee a sum of money to cover for on-cost and profit. However, the incentive may be applied by having a fluctuating fee based on an agreed and priced bill of quantities, and if the total cost of the work is greater than the estimated cost the fee is reduced by, say, 5 per cent or, indeed, by any other relatively small percentage. On the other hand, if the cost of the work turns out to be less than the estimated cost the saving is shared between the contractor and employer on some basis, which may even be 50 per cent to either side.

Often a target time with incentives for completion on time is also adopted for contracts subject to target costs. Unlike in contracts where a predetermined sum per week is arrived at, as liquidated damages payable by the contractor for non-completion of the works in due time, the twin-target contract offers a monetary incentive to the contractor for completion ahead of the predetermined date, while completion after this date attracts a sum of money due to the employer from the contractor. Again as in the case of target costs, it is usual for the rate at which the

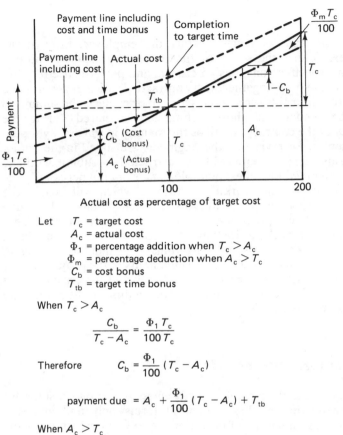

Let T_c = target cost
      A_c = actual cost
      Φ_1 = percentage addition when $T_c > A_c$
      Φ_m = percentage deduction when $A_c > T_c$
      C_b = cost bonus
      T_{tb} = target time bonus

When $T_c > A_c$

$$\frac{C_b}{T_c - A_c} = \frac{\Phi_1 T_c}{100 T_c}$$

Therefore $C_b = \dfrac{\Phi_1}{100}(T_c - A_c)$

payment due $= A_c + \dfrac{\Phi_1}{100}(T_c - A_c) + T_{tb}$

When $A_c > T_c$

$$\frac{C_b}{A_c - T_c} = \frac{\Phi_1 T_c}{100 T_c}$$

Therefore $C_b = \dfrac{\Phi_m}{100}(A_c - T_c)$

payment due $= A_c \dfrac{-\Phi_m}{100}(A_c - T_c) + T_{tb}$

Figure 3.1. Generic case for a cost-reimbursable contract with twin targets

contractor is rewarded for completion ahead of time to be greater than the rate at which the contractor is made liable to reimburse the employer for non-completion by the set date.

The basis of payment on a twin-target contract having both a predetermined cost and time for completion may be expressed in a generic form, as shown in Figure 3.1.

Design and construct contracts

In a design and construct contract the employer, through the engineer, defines the scope of the project by means of outline drawings, specifications, design criteria and performance requirements. These data, together with contractual, pricing, general and special requirements such as the scope of drawings to be submitted with the tender and those submissions required during the currency of the contract as well as the manner in which they should be presented for review by the engineer, are stated in the tender documents. The tenderer will be instructed to submit with the tender, *inter alia*, the fullest possible details but not necessarily the complete design and working drawings, which will enable the tenderer's proposals to be appraised. Such details will include pricing schedules; programmes for the development of design, preparation of working drawings and construction of the works; and design calculations and drawings pertaining to the civil, structural, architectural and E & M elements of the project, where appropriate. The tender period for a design and construct contract varies between three and six months, depending on the nature and complexity of the project.

All-in or turnkey contracts

We now come to the type of contract where the employer states his requirements in broad and general terms only and invites a contractor to submit an all-in bid or turnkey bid, which is for the provision of the whole of the work including its design, construction and, if required, the commissioning operation and maintenance for a limited period of the project. The employer may also wish in such a contract to be financed until the project is earning money. This type of contract is well known and well used in the USA and elsewhere for the establishment of nuclear power stations, chemical process plant and other specialized high-technology manufacturing process projects, the principal examples of which in the UK are the nuclear power stations undertaken by the Central Electricity Generating Board.

In order to undertake turnkey contracts for highly specialized projects, where in-house capability in all the disciplines is not available, manufacturing and contracting firms have formed themselves into consortia, with the design element being carried out by consulting engineers retained by them.

Where the promoter's own organization does not have the technical resources to provide tender documents, performance specifications, design criteria and other parameters for the project it is wise to obtain advice from a consulting engineer well versed on these aspects. Such an engineer will also be in a position to advise the promoter on the selection of a contractor and help to steer the contract until the plant is fully operational. A prerequisite of a turnkey contract is that the responsibility for the design, manufacture, construction and the performance of the plant should be made to rest with the turnkey contractor.

Management contracts

This type of contract, which was first introduced on building complexes in the late 1960s, has now spread to a wide range of projects. In its basic form, management contracting is the term used when a contractor experienced in carrying out multi-disciplinary projects is entrusted with the overall management of several sub-contractors, more appropriately called construction contractors. The management contractor first provides his expertise on planning a project to the employer and his professional advisers. When the basic project planning is complete he is appointed to coordinate and control the construction contractors, who are selected by the employer on the recommendation of his professional advisers: this will be following the assessment of tenders called on a competitive basis.

The management contractor does not normally undertake any construction work on the site, and his fee is based on a percentage of the estimated value of the project plus his own costs for services rendered by him until he is appointed to manage the contract. The engineer's role on management contracts is sensibly similar to that on traditional contracts: i.e. the responsibility for design, supervision of construction, issuance of variation orders, certification of payment plus the continuous audit of project accounts.

Several variations on the basic form of management contract have been developed. These include design and management contracts; construction management contracts; and the provision of contract management services. It is essential that an employer should analyse precisely the services offered under this form of contract to ensure that the responsibilities of the parties are clearly defined.

Work by direct labour

The employer may decide that he is to carry out works by means of what is known as direct labour. In effect this means that no contract exists but that he carries out all the work on his own account, employing the staff and labour and using his own or acquiring or hiring plant and buying the materials, both temporary and permanent, that are required.

This used to be a not-uncommon method of carrying out work for large projects, but, in general, nowadays this is limited to undertakings carrying out works of small account and, of course, maintenance work in this manner.

This method of carrying out work means that the employer can design and construct exactly what he requires without any external interference from a contractor or a consulting engineer but it has, of course, the great disadvantage that there is little incentive to complete the work at the least cost and in the shortest possible time.

Overall parameters

In considering the various types of contracts available it should be borne in mind that each type has its advantages and disadvantages, depending on the category of work and the importance attached by the employer in retaining the following three parameters:

1. Financial foreknowledge;
2. Technical flexibility;
3. Risk.

The level to which each of these parameters is retained by the employer determines the type of contract most suited to his needs. The first parameter is inversely related to the other two, and this relationship is illustrated in Figure 3.2.

The three parameters discussed above may be expressed in the following terms:

Financial foreknowledge or financial certainty is the degree to which the financial commitment of the work to be undertaken is available to the employer.
Technical flexibility is the degree of freedom the employer has to vary the works at any stage from conception to completion to meet his varying needs or to react to varying technical requirements.

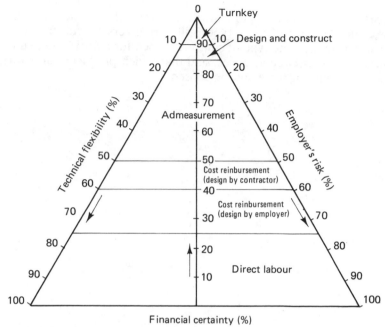

Figure 3.2. Risk parameters of types of contract

Risk is the cost of unforeseen circumstances which may arise during the course of the contract and which must either be allowed for in the contractor's price or be borne by the employer if the work is to be satisfactorily completed.

At one end of the scale lies the turnkey or all-in contract, where the financial risk to the employer as well as the technical flexibility that he has are at a minimum. The degree of financial foreknowledge is, of course, the highest on this type of contract. In taking on the risk the contractor has to increase his offer. Likewise, the low technical flexibility will require the employer to define his project fully at the start: any changes that occur in this respect could prove very costly.

At the other end of the scale lies work carried out by direct labour. On work carried out in this manner the employer takes all the risk, paying for an eventuality that may arise and retaining complete flexibility on his requirements until completion. Under these circumstances he must sacrifice financial foreknowledge, since, until the last item is completed, the cost of the work is unknown.

Between the limits of the turnkey contract and direct labour work lie the other types of contract and the risk and the technical flexibility of each of these types will vary inversely with financial foreknowledge. The type of contract chosen for a project depends on a variety of circumstances, and that which produces value for money should in the event be chosen.

Chapter 4

Contract risk and contract responsibility

At the outset a definition should be given as to what is meant by *contract risk* and *contract responsibility*.

In civil engineering contracts – and indeed in other types of contract – the successful bidder, who is translated into the contractor, has undertaken to carry out certain works. It is he who is at risk (or not) financially in everything he constructs.

To be more specific, a contract risk is something for which the contractor takes full financial responsibility and a contract responsibility is anything which the contractor has to carry out, although in doing so he is not at risk financially.

In any project the tender documents and drawings will contain what may be termed the financial risk element in many facets of the contract. It follows that each contract has to be considered within its particular context.

In consideration of a civil engineering contract and, to be more explicit, one using the ICE Conditions of Contract, the following are examples of contract risks:

1. Obtaining the required labour, material and plant with which to carry out the work.
2. Weather.
3. Compliance with the requisite standards of materials and workmanship.
4. Accidents.
5. Mistakes, such as errors in setting out.

Again, the following are examples of contract responsibilities:

1. The carrying out of dayworks.
2. The carrying out of works connected with clause 12 in cases where that clause has been successfully invoked by the contractor.
3. Items such as provision of pumps for the dewatering of foundations, where keeping the work clear of water is not written into the relevant item in the priced bill of quantities.

It will be seen therefore that everything which a contractor carries out in the construction of a civil engineering contract is either a contract risk or a contract responsibility, and because of this, whether it be in rate fixing, in issuing instructions, in the resolution of claims or in making decisions, the engineer must have clearly in mind into which category a particular item falls.

Let us now see how this whole question of contract risk and contract responsibility is dealt with in practice.

To take one example, a contractor may claim because he has not been able to supply a sufficient number of carpenters: or he may claim that portland cement deliveries have failed to materialize. Neither claim is valid, since both elements, by definition, are contract risks.

Let us next take an example of a condition where, due to a failure on the part of the employer or the engineer or for some extraneous reason, the contractor has not been provided with working drawings in due time. The drawings are, of course, something which the contractor does not carry out and which is, therefore, neither a contract risk nor a contract responsibility. However, if the claim is valid, it is obviously not the fault of the contractor, and he should be recompensed for any loss – if any exists – that he has suffered due to this delay.

As our next example we may take an excavation item where neither the engineer nor the contractor can forecast how much water is to be pumped, and it would be fair if the engineer were to lay down in the item on excavation that it shall include pumping up to, say, 30 litres/second of water. In this case the provision of pumps and their proper function is a contract risk up to this limit, but any water over 30 litres/second is a contract responsibility which is paid for separately either through a daywork item or through a special item in the bill of quantities. In the latter case, of course, there is some element of contract risk involved if the contractor was asked to insert a rate-only item.

We have tried to demonstrate the importance of appreciating the contract risk and contract responsibility elements in a civil engineering contract and this involves the arbitrator in cases where a dispute has led to arbitration proceedings. The arbitrator in effect will be reviewing a decision of the engineer, albeit with data differing from those available to the engineer with which to review that decision. It follows that the award by the arbitrator is in effect going over the ground of the engineer's decision and superimposing his conclusion as to the correctitude or otherwise of that decision. It may not be irrelevant to point out that proper consideration of contract risk and contract responsibility leads to better documentation of a civil engineering contract. Here we are specifically considering the question of imponderables, and it is very often better for the engineer to remove, so far as is possible and practical, items from the category of the contract risk to that of the contract responsibility. This is illustrated in the following example.

Let us first take a rock tunnel where, in spite of a proper site investigation, one cannot forecast exactly how the rock will behave on excavation. In former times the provision of temporary supports was always put into the contract risk category – that is, the contractor's price would include temporary supports as might be required (the supports are loaded temporarily but are usually left in place). Now, if the contractor assumes that, say, one-third of the tunnel has to be supported and in the event no supports are required, then he will gain financially to the detriment of the employer. If, on the other hand, the whole of the tunnel has to be supported, he will lose money and obviously try to recoup this by invoking clause 12 of the ICE Conditions of Contract, on the grounds that an experienced contractor could not reasonably have anticipated the need to line the whole of the tunnel with temporary supports. From the foregoing it will be seen that it is better to write the temporary supports into the contract as a contract responsibility; in other words, there would be an item in the priced bill of quantities for provision of the supports as a rated item subject, of course, to admeasurement. The contractor would be paid for just what was used; delays arising from the provision of the supports will still remain a contract risk.

Let us take as our next example a caisson, a tunnel or any other structure which requires the use of compressed air working. In the past, the contractor was often asked to gamble on whether compressed air was required or not. To no small degree, this is an imponderable, and it is better to write an item into the priced bill of quantities for the provision of compressed air plant and its operation and to have extra over items for working in compressed air. Again, the amount of compressed air to be used may be limited in the contract risk aspect by giving a quantity of air which has to be provided and then quoting an additional item for compressed air quantities over that amount.

Disputes on such matters often finish in the hands of an arbitrator who frequently has a difficult task, as well as insufficient data. Claims and disputes arising from invoking clause 12 are numerous and often particularly difficult to resolve. In the case where disputes are not easy to settle, the engineer is forced to make a decision very near the boundary between a positive or negative answer. In all cases, true appreciation of the contract risk and contract responsibility is essential both from the point of view of the engineer and the arbitrator and, indeed, it must be clear in the mind of the contractor. Notwithstanding this, one has obvious sympathy with all those concerned in borderline cases. It is certainly an advantage that in the latest ICE Conditions of Contract arbitration may now take place immediately a clause 12 decision has been given by the engineer against the claim. In this way there is often a good chance of the arbitrator being able to see and to appreciate the situation as it exists and not, as in so many cases in the past, have to make his award largely on hearsay.

Chapter 5

Contract documents and their application

Contract documents are those in which the scope and the requirements of a project are comprehensively laid down. In these documents the obligations and responsibilities of the parties to the contract as well as the engineer's powers, duties and functions which flow from such a contract are defined. The engineer is not generally a party to the contract and therefore has no contractual rights or obligations under it.

The tender documents issued to prospective tenderers are in effect pro-forma contract documents. When the blanks in these documents are filled in by a tenderer, and an offer made thereon is accepted by the employer, these documents together with any amendments, modifications and amplifications that are agreed upon before the final acceptance of the tender become the contract documents governing the performance of the contract.

The types of contract normally adopted for civil engineering works were described in Chapter 3. These fell into five main types, namely, admeasurement, lump-sum, cost/reimbursement; design and construct, and all-in or turnkey. Whichever type of contract is adopted, the obligations and the responsibilities of the parties to the contract have to be defined in the documents in a clear, unambiguous form so as to enable the contractor to assess his risks and responsibilities and price for the works to be carried out by him in a realistic manner. The smooth operation of a contract depends largely on the clarity and adequacy of the contract documents.

The contract documents adopted for civil engineering construction on an admeasurement basis (the most common type in the UK) comprise the following elements:

1. Instructions to tenderers
2. Form of tender and appendix
3. Form of agreement
4. Form of bond
5. Conditions of contract

6.   Special conditions of contract
7.   Specification
8.   Schedule of basic prices
9.   Bill of quantities
10.   Daywork schedule
11.   Contract drawings.

If the type of contract is not on an admeasurement basis but falls into the category of a lump-sum, cost/reimbursement, design and construct or a turnkey type then all the elements set out above are not required.

A lump-sum contract does not normally contain a bill of quantities, nor a schedule of basic rates. However, in order to cope with possible variations to the works covered by the contract, but not required to be executed within the lump-sum, a complete bill of quantities is sometimes called for from the contractor to determine the value of such a variation; the value so determined can then be added to or deducted from the lump-sum. Other ways of providing for variations in a lump-sum contract are by obtaining at the tender stage a schedule of unit rates for the different elements of work or for labour, plant and materials on a daywork basis. In any event, if any variations are envisaged in a lump-sum contract provision should be made in the Tender Enquiry to deal with such variations, the absence of which will invariably lead to very high costs even though the variations required may be small.

In cost/reimbursement contracts generally a basic schedule of rates and a bill of quantities are not incorporated. However, if an incentive to the contractor to restrict the capital cost is provided within the contract by the application of a target cost then it is usual to incorporate a bill of quantities for this purpose.

For a design and construct contract, detail drawings and specifications are not provided. However, outline drawings, design criteria and performance requirements are laid down. The scope of drawings and documents to be submitted with the tender and those submissions required during the currency of the contract are also called for in the tender documents.

In an all-in or a turnkey project the design of the permanent work is an added responsibility for the contractor. It follows that detailed drawings and bill of quantities are therefore not prepared by the engineer for such contracts. However, it is necessary to lay down the scope of work to be carried out in unequivocal terms in the tender documents to enable the contractor fully to interpret the employer's intentions. The specification for a turnkey project should contain sufficient layout drawings expressing the employer's requirements; also it is most important to include a comprehensive specification setting out the design criteria to be adopted together with such documents relating to performance and operational requirements to ensure that the desired end product is achieved.

When mechanical equipment, heating and ventilation equipment, electrical installations and the like are to be incorporated into the works it is usual to issue schedules of technical requirements which are design parameters and performance criteria related to these items. It is also necessary to obtain from the tenderer schedules of technical particulars in which the items offered are fully described.

A schedule of exclusions, if any, should also be called for from the tenderer so that the content of the contract for which prices are offered is expressly defined.

The elements that comprise the tender documents for an admeasurement contract may be expanded as set out below.

Instructions to tenderers

The Instructions to Tenderers is usually the first section included in the bound volumes of the tender documents. The purpose of this section is to ensure that all tenders are properly prepared and delivered so that they may be evaluated on an equal basis. These instructions will vary on every project but some of the more usual and important items to be included would be as follows:

1. Instructions on filling in tender forms.
2. Data to accompany tenders:
   (i)   Documents to be submitted with tender.
   (ii) Programme.
   (iii) Memorandum of procedure.
   (iv) Letter of capacity.
   (v)   Schedules to accompany tender.
3. Delivery of tenders:
   (i)   The original and the number of copies to be submitted.
   (ii) The name or the designation of the official and the organization or organizations to whom the tenders are to be submitted.
   (iii) The place, date and due time for delivery of tenders.
4. Letter of capacity. This is called for only if the qualifications of the tenderers have not been exhaustively assessed prior to addressing the Enquiry. The letter of capacity will include:

(i)   Proof of execution of projects of a similar scale and kind to that contemplated in the Enquiry.

(ii) Qualifications of the personnel proposed to be employed on the project.

(iii) Plant resources of the firm.

(iv) Management and financial resources of the organization.

(v)   Legal status of the tenderer.

5. Admissibility of alternative designs, instructions on whether tenders on alternative designs will be considered and, if so, the conditions under which they may be submitted.

6. Validity and price variations:

(i)   The period of the validity of the tender.

(ii) Whether the contract is to be on a fixed-price basis or whether the contract price is to be subject to cost fluctuations.

7. Special conditions: notes on access and any unusual site conditions should be stated.

8. Instructions on completion of the bill of quantities, schedule of rates, daywork schedule where applicable and the requirement for tender bond, if any, should be stated.

Form of Tender

The Form of Tender is the part of the document the tenderer is required to fill in to make an offer to execute the works. For example, the Institution of Civil Engineers (ICE) Form of Tender is so worded that in completing the form the tenderer is deemed to have taken cognizance of the following:

1. Examined the drawings, conditions of contract, specification and bill of quantities for the construction of the works named in the form.

2. That for such sum as may be ascertained from the conditions of contract he would construct, complete and maintain the works in conformity with the drawings, conditions of contract, specification and bill of quantities.

3. The work shall be completed within the time stated in the appendix to the Form of Tender.

4. On acceptance of the tender that the contractor, if required, shall provide a performance bond in a sum not exceeding 10 per cent of the tender sum.

5. That unless and until a formal agreement is prepared and executed his tender, together with the employer's written acceptance thereof, shall constitute a binding contract between the employer and contractor.

The appendix to the Form of Tender is part of the tender. In the appendix, figures which need to be inserted under the various clauses in the conditions of contract are filled in.

Form of Agreement

The Form of Agreement is the formal document confirming the existence of a contract between the employer and the contractor. In the United Kingdom this form is generally not used as the acceptance of the tender in itself forms a contract. However, certain local authorities and statutory undertakings insist that the formal agreement is completed; indeed these bodies under certain circumstances are required to execute the agreement under seal in terms of their acts of creation or rules that are laid down for the conduct of their business. In the execution of large contracts carried out abroad formal agreements are often executed.

Form of Bond

The Form of Bond is the legal document by which the contractor gives effect to the requirement of providing a performance bond for the contract in a sum usually not exceeding 10 per cent of the tender sum.

The contractor has the option of providing two good and sufficient sureties or to obtain the guarantee of an insurance company or bank to be jointly and severally bound with the contractor for the sum required for this purpose by the employer.

The bond must be in writing and should be under seal. It provides that if the contractor completes and maintains the works the covenant becomes void. Sureties are normally released by major contract variations. The employer must satisfy himself regarding the status of sureties which are normally provided by insurance companies, who charge the contractor on a percentage basis. A bond may add a further one per cent to the contract price. Frequently a performance bond is not called for in civil engineering contracts operated in the United Kingdom. It should be noted that the employer under the conditions of contract is covered against the default of the contractor by retention monies, plant on site, and also the lag in payment that arises after the contractor has physically carried out the work. However, in overseas contracts it is quite common for a performance bond to be called for in almost every contract.

Conditions of Contract

The Conditions of Contract incorporated into contract documents are generally one of the standard forms of contract such as the Institution of Civil Engineers (ICE) Conditions of Contract, GC/Works/1 Conditions of Contract for Overseas Works mainly of Civil Engineering Construction prepared by the Association of Consulting Engineers and the International Conditions of Contract (FIDIC) or modifications thereof.

Special conditions of contract

The special conditions of contract are essentially those which have to be drafted to suit a particular situation. They are matters for which clauses are not included in the standard conditions of contract such as the ICE, Overseas, FIDIC or such other conditions of contract prepared for general application. Both the Overseas and the FIDIC documents set out guidance notes for the preparation of special conditions of contract; these aspects will be described in some detail in the next chapter.

Specification

The scope of works together with the general and technical requirements of a project is described in the specification. This element of the contract documents is dealt with in Chapter 7.

The Schedule of Basic Prices, the Bill of Quantities and the Daywork Schedule are the three elements of the contract documents which are concerned with the quantification and pricing of the items of work carried out under the contract. These are dealt with in Chapter 8.

Contract Drawings

The Contract Drawings are the drawings on which the works are to be carried out and on which the contract price is based. The ICE Conditions of Contract defines 'drawings' as those referred to in the specification and any modifications of such drawings approved in writing by the engineer and such other drawings as may from time to time be furnished or approved in writing by the engineer.

This, in effect, imposes an obligation on the contractor to carry out the works not only in accordance with the drawings issued at the time of tender, but also with those issued during the currency of the contract. Of course, the contractor is entitled to reimbursement of additional costs for works carried on modified drawings to the extent that the contract provides for reimbursement in respect of such modifications by means of admeasurement and, if required, by rate fixing.

Sub-contracts

The subject of sub-contracts is discussed in this chapter because it often forms a part of tender documents issued to prospective contractors for pricing.

With the increasing amount of specialist services and products that are required for civil engineering projects the employment of sub-contractors for such services by contractors is now widespread. Depending upon the nature and the degree of urgency of providing the specialist service in relation to the overall programme of a project, frequently 'sub-contractors' are chosen by the engineer before the main contract is let. These sub-contractors as well as those required by the engineer to be employed by the contractor during the currency of the contract become nominated sub-contractors under the main contract. The term main contract is used here to denote the contract between the employer and the contractor.

A distinction must be drawn between a nominated sub-contractor and what is generally known as a 'domestic sub-contractor'. The latter is employed by the contractor entirely at his own volition, provided the engineer consents to such employment. A domestic sub-contract could be let by a contractor for any discrete part of the works. This could be for the construction of roads, the supply and installation of small power and lighting or for the supply and erection of structural steelwork. Such sub-contracts when let by the contractor are deemed to be work carried out by the contractor for which he is wholly liable and responsible.

When sub-contractors are nominated by the engineer the contractor is still not relieved from any liabilities or obligations in respect of the work carried out by such nominated sub-contractors. However, under standard conditions of contract and particularly the ICE Conditions of Contract there are certain prerequisites that have to be conformed to before the contractor can be held to be wholly responsible for all the liabilities and obligations arising from such sub-contracts.

The ICE Conditions cover the subject of Provisional and Prime Cost Sums and Nominated Sub-Contracts at clauses 58, 59A and 59B. Before dealing with the principles of nominated sub-contracting it may be as well to define these items as set out in the ICE Conditions.

Provisional Sum

'Provisional Sum' means a sum included in the contract and so designated for the execution of work or the supply of goods, materials or services or for contingencies, which sum may be used in whole or in part or not at all at the direction and discretion of the engineer.

Prime Cost Item

'Prime Cost (PC) Item' means an item in the contract which contains (either wholly or in part) a sum referred to as Prime Cost (PC) which will be used in the execution of work or for the supply of goods, materials or services for the works.

Nominated Sub-contractors

All specialists, merchants, tradesmen and others nominated in the contract for a Prime Cost Item or ordered by the engineer to be employed by the contractor in accordance with other provisions in the contract for the execution of any work or the supply of any goods or materials or services are referred to in the contract as 'nominated sub-contractors'.

The basis of sub-contracting amounts to the sub-letting of the physical construction of the works only. The contractor remains fully liable to the employer under clause 4 of the conditions of contract, where it is specifically laid down that the contractor shall be responsible for 'the acts, defaults, or neglects of any sub-contractor, his agents, servants or workmen as fully as if they were the acts defaults or neglects of the contractor, his agents, servants or workmen . . .'

On the other hand, the contractor is not under any obligation to enter into a nominated sub-contract with a sub-contractor against whom the contractor can raise a reasonable objection. Likewise, the contractor can refuse to enter into a contract with a nominated sub-contractor who is not prepared to undertake towards the contractor such obligations and liabilities as will enable the contractor to discharge his own obligations and liabilities towards the employer under the contract.

If the contractor is not obliged to enter into the sub-contract on the above grounds it is open to the engineer to act as follows:

1. Nominate an alternative sub-contractor.
2. Vary the works or omit them with compensation as provided for under the contract.
3. Direct the contractor to enter into a sub-contract and at the same time, among other things, relieve the contractor from the discharge of his obligations and liabilities to the extent the sub-contract terms specified by the engineer are inconsistent with the discharge of the contract.

The sub-contract is entered into between the contractor and sub-contractor, and the employer has no contractual rights under the sub-contract against the sub-contractor. Likewise, the sub-contractor cannot sue the employer direct for payment or extra expenses or damages. In the engineer's communications to sub-contractors the privity of the contract between the contractor and the sub-contractor must be clearly maintained, and on no account should the engineer create a direct contract on behalf of the employer with a sub-contractor, in which case the whole object of the sub-contracting system from the employer's point of view would be vitiated.

Placing of sub-contracts

Nominated sub-contracts are placed where specialist services are required to be incorporated into the main contract. In a civil engineering project such specialized services, which include piling, precast concrete components, heating and ventilating services, electrical services, etc., may be carried out on the basis of nominated sub-contracts. The engineer, where appropriate, prepares bills of quantities, specifications and data on required performance, and calls for tenders under the terms of the contract from a selected list of specialist contractors. The employer, on the advice of the engineer, or the engineer chooses a sub-contractor whose work is incorporated into the main contract.

In order to accommodate a sub-contract the bill of quantities of the main contract contains a Prime Cost item or a Provisional Sum for the sub-contract work envisaged. In addition, usually in the

'General Items' bill the contractor is required to quote the percentage required by him for attendance, charges and profit on Provisional Sums set aside for sub-contract work.

After the initial negotiations are made by the engineer with the sub-contractor, the contractor is instructed to place an order with the sub-contractor under the provisions of clause 59 of the Conditions of Contract. Alternatively it is also established practice for the negotiations with a sub-contractor to be entirely conducted between the contractor and sub-contractor and for the sub-contract to be ratified by the engineer.

Apart from nominated specialist firms the engineer may nominate other sub-contractors for the supply of particular goods and fittings.

As the contractor is responsible for all acts and defaults of his sub-contractors, his agreement with them should call for like obligations and liabilities from the sub-contractor that he would have to bear to the employer. The following, among other things, are matters which should be covered in the sub-contract agreement:

1. Description of work, method of execution, access to work and working areas allowed.
2. Terms of payment, date of completion, provision regarding extras and variations and the extent of cover provided under the contractor's insurance.
3. Sub-contractor to indemnify the contractor for non-completion of the work in time, liquidated damages, etc.
4. Provision for arbitration in the event of a dispute.

The conditions of the main contract do not normally refer specifically to the form in which the sub-contract has to be entered into by the contractor. As a result, difficulties can be caused by the failure of contractors and sub-contractors to complete a proper sub-contract before the work is started.

The Federation of Civil Engineering Contractors in the United Kingdom publishes a form of sub-contract for use in conjunction with the ICE General Conditions of Contract which the contractor should be encouraged to use wherever practicable. This form was first published in June 1973 for use with the ICE fourth edition (1955). It was subsequently used without amendment with the fifth edition of the ICE form and, indeed, with the ICE Overseas (Civil) Conditions and the FIDIC (Civil) forms, until it was marginally revised in September 1984.

This form comprises an Agreement, 20 clauses and five schedules, and the clauses cover the following elements

Clause 1: Definitions
Clause 2: General Requirements
Clause 3: Main Contract
Clause 4: Contractor's Facilities
Clause 5: Site Working and Access
Clause 6: Commencement and Completion
Clause 7: Instructions and Decisions
Clause 8: Variations
Clause 9: Valuation of Variations
Clause 10: Notices and Claims
Clause 11: Property in Materials and Plant
Clause 12: Indemnities
Clause 13: Maintenance and Defects
Clause 14: Insurances
Clause 15: Payment
Clause 16: Determination of the Main Contract
Clause 17: Sub-contractor's Default
Clause 18: Disputes
Clause 19: Value Added Tax
Clause 20: Law of Sub-Contract

The first schedule is for the enumeration of the particulars of the main contract: these are the parties to the contract; the date; a brief description of the main works; the dates on which the contractor will be submitting statements (payment) to the employer; and the minimum amount of interim certificates under the main contract.

The second schedule is for the setting out of the further documents that form part of the sub-contract, a description of the sub-contract works and the fluctuation provisions (if any) that affect payment under the sub-contract agreement.

The third schedule provides for the inclusion of the price, the percentage of retention on work carried out, the percentage of retention on materials on-site, the limit of retention and the period for completion.

The fourth schedule comprises two parts. Part I is for the enumeration of common facilities that will be provided to the sub-contractor in the way of constructional plant and others with the terms and conditions attached to such provision. Part II is for a similar statement of exclusive facilities that are to be provided to the sub-contractor.

The fifth schedule is for a statement of the insurances that are to be effected by the sub-contractor and those that will be effected by the contractor for the benefit of the sub-contractor.

Chapter 6

General conditions of contract

The General Conditions of Contract define the terms under which the works are to be executed and maintained and set forth the obligations and liabilities of the parties to the contract. The powers of the engineer, the terms of payment and the manner in which a dispute under the contract is to be settled are a few of the many aspects covered by these conditions.

Conditions of contract of various forms are used in the execution of civil engineering works. In the United Kingdom the form used is invariably the Institution of Civil Engineers Conditions of Contract, although at times these are modified to suit the needs of the particular organization that uses them.

The ICE Conditions of Contract were first issued in 1945; the current document is the fifth edition published in June 1973, revised in January 1979 and reprinted in January 1986 to incorporate amendments issued since January 1979. The ICE Conditions are designed for use on admeasurement contracts; this type of contract is adopted in the majority of civil engineering projects in the United Kingdom.

Before proceeding to the details of the latest edition of the ICE Conditions of Contract it would not be inappropriate to review briefly the whole question of general conditions of contract in the field of civil engineering.

Before the issue of ICE Conditions of Contract, public undertakings and local authorities – and indeed consulting engineers – all had their own general conditions of contract which, as may be imagined, varied greatly. Indeed, one major authority under the auspices of the government has until fairly recently been using its own conditions which date back to the days of Samuel Pepys. All this meant that both in the preparation of tender documents and in the submission of tenders all the individual general conditions of contract would have to be studied on their own, and this was time consuming.

In the construction stages of any project, all concerned had to interpret general conditions of contract with which they were not necessarily familiar.

The advent of a standard form of general conditions of contract in 1945 was, therefore, an excellent conception, although to some extent it contributed to the decrease in the power of the engineer, which had been preserved since the end of the First World War, if not before. It is certainly not irrelevant to mention this matter, since it may be stated that the first issue of the ICE Conditions of Contract took a further step in this direction, and it is also to be regretted that the latest Conditions have continued this trend. It may be said that this trend is not a good one and has not been of benefit to either the employer or the contractor.

At all events, the present situation is that a majority of undertakings, local authorities and consulting engineers have adopted the ICE Conditions of Contract in an increasing measure since their first issue, although there are a few of them who are still operating under different conditions of contract.

The Institution of Civil Engineers, the Association of Consulting Engineers and the Federation of Civil Engineering Contractors have, as sponsoring authorities, approved these conditions of contract for all works of civil engineering construction. The interests of these bodies as well as those of the employers were represented on the committee set up to draft the conditions.

The fifth edition of the ICE Conditions of Contract contains 72 clauses, the first being 'Definition and Interpretation' and the seventy-second is for stipulating any special conditions which it is desired to incorporate into the conditions of contract. Such conditions are numbered consecutively from the last standard condition set out in the document. Included with the conditions of contract are the form of tender together with the appendix thereto, the form of agreement and the form of bond.

The clauses in the conditions of contract are grouped under the following twenty-five heads:

Definitions and interpretation
Engineer's representative
Assignment and sub-letting
Contract documents
General obligations
Labour
Workmanship and materials
Commencement time and delays
Liquidated damages and limitation of damages for delayed completion
Completion certificate
Maintenance and defects

Alterations, additions and omissions
Property in materials and plant
Measurement
Provisional and prime cost sums and nominated sub-contracts
Certificates and payment
Remedies and powers
Frustration
War clause
Settlement of disputes
Application to Scotland
Notices
Tax matters
Metrication
Special conditions.

The several clauses in the conditions of contract could be said to relate to five distinct facets of the construction contract. They are policy in relation to the contract; execution of the works; time within which the works are to be executed; payment for the work carried out; and default and disputes arising from the contract.

All clauses in the conditions of contract are relevant to the works and binding on the parties when they are incorporated into the construction contract. Certain aspects of the subject matter of some of the clauses are set out below.

Policy in relation to the contract

Clause 1
This clause deals with definitions and accordingly defines the terms that are to be used in the clauses that follow. Points of particular significance are that 1(1)(h) defines the 'Tender Total' as the total of the priced bill of quantities at the date of acceptance of the tender and that under 1(1)(i) the 'Contract Price' is defined as the sum to be ascertained and paid in accordance with the provisions of the contract in respect of the works. This definition together with some of the clauses that follow make it clear that the works are to be valued on an admeasurement basis. Also it is worth noting that 1(5) defines 'Cost' to include overhead costs whether on- or off-site except where the contrary is expressly stated.

Clause 4
This requires the contractor to obtain the engineer's written consent before sub-letting any part of the works.

Clause 5
This clause gives equal weight to all documents that form the contract. The engineer is empowered to explain and adjust any discrepancies or ambiguities which may arise and instruct the contractor accordingly.

Clause 9
The requirement for a formal contract agreement to be executed should the contractor be called upon to do so is stipulated in this clause.

Clause 10
The requirement for a performance bond to be provided if such an undertaking has been given in the tender is stipulated in this clause.

Clause 51
This clause empowers the engineer to order any variation to any part of the works which in the engineer's opinion is necessary for the completion of the works or is desirable for the satisfactory completion and functioning of the works.

Clause 67
This clause requires that the contract shall be construed and operated as a Scottish contract in all respects should the works be situated in Scotland.

Clause 68
The places for the service of notices from the employer to the contractor and vice versa are stipulated in this clause.

Execution of the works

Clause 1
Several terms in connection with the execution of the works such as 'Engineer's Representative', 'Specification', 'Drawings', 'Permanent Works', 'Temporary Works', 'Works' and 'Site' are defined in this clause. Of particular significance is that at 1(1)(f) the specification is defined as that referred to in the tender and any modifications or additions made to this from time to time in writing by the engineer; the modifications to drawings and the issue of additional drawings by the engineer during the currency of the contract are treated at 1(1)(g) in a like manner.

Clause 2
This clause deals with the engineer's representative. The powers, duties and functions that flow from this clause to the engineer's representative are dealt with in Chapter 11.

Clause 8
The contractor's general responsibilities are set out in this clause. The responsibilities of the contractor towards the safety of site operations are stipulated at sub-clause (2) of this clause.

Clause 11
By this clause the contractor is deemed to have inspected the site and its surroundings and satisfied himself before submitting the tender on the nature of the ground and sub-soil and all other circumstances affecting his tender. Under 11(2) the contractor has to ensure that his tendered rates and prices are correct and sufficient to fulfil all his obligations under the contract.

Clause 12
This clause provides a means for the contractor to obtain additional payment and an extension of time for completion should he encounter adverse physical conditions or artificial obstructions during the execution of the works which he as an experienced contractor could not have reasonably foreseen. The procedure for making claims under this clause and the procedure the engineer is to follow in respect of such claims are set out in this clause.

Clause 13
The engineer's power to control the works stems from this clause to a considerable degree. The clause, *inter alia*, empowers the engineer to instruct and direct the works on any matter connected therewith provided it is legally and physically possible.

Clause 14
This clause is concerned with the programme of works, method statements and temporary works proposed by the contractor to carry out the works. The manner is which these proposals are to be submitted to the engineer for his approval or consent and other action to be taken in this regard are covered under this clause. A point to note is that the contractor's responsibility for this work is unaffected by the approval or consent given by the engineer.

Clause 17
This clause deals with the setting-out required for the permanent works. The responsibility for setting-out remains with the contractor even though it is approved by the engineer or his representative. The cost of rectification of any errors is to be borne by the contractor unless it is due to incorrect data supplied by the engineer or his representative.

Clause 20
This clause sets out the contractor's responsibilities for the care of the works and is specific on the period for which he is so responsible. Sub-clause (3) stipulates the 'excepted risks' which are normally uninsurable. Under this clause the contractor is responsible for the care of the works whether or not they are insured.

Clauses 21–25
These clauses cover the insurance requirements and the indemnities the contractor has to extend to the employer and vice versa. Insurance in the context of civil engineering contracts is dealt with in Chapter 9.

Clause 30
This clause deals with the avoidance of damage to highways or bridges from being subjected to extraordinary traffic during transport of constructional plant, materials and fabricated articles to site. The action that should be taken including strengthening of bridges and improvements to roads together with the manner in which damages, costs, charges and expenses arising from extraordinary traffic is also dealt with in this clause.

Clause 31
This clause empowers the engineer to direct the contractor to afford reasonable facilities for other contractors employed by the employer and authorized persons who are engaged on or near the site.

Clause 33
This clause requires the contractor to leave the whole of the site and permanent works in a clean and workmanlike manner on completion of the works.

Clause 34
This clause sets out the requirements of the Fair Wages Resolution Act 1946 concerning wages and conditions that the contractor is required to observe. This also stipulates that the requirements of the Civil Engineering Construction Conciliation Board for Great Britain should be complied with.

Clause 35
This clause empowers the engineer to call for plant and labour returns from the contractor or from any of his sub-contractors.

Clause 36
The quality of materials and workmanship to be provided for the works together with the contractor's obligations in respect of sampling and testing are covered by this clause. A point to note is that the cost of tests ordered by the engineer that are not required by the contract are to be borne by the contractor if the test fails but otherwise by the employer.

Clause 37
This clause gives the right to the engineer or any person authorized by him to have access to the site and the contractor's places of work.

Clause 38
This clause requires the contractor to obtain the engineer's approval before covering up any work. The clause also empowers the engineer to direct the contractor to uncover any part of the work and should such work be defective the cost of uncovering and making good is to be borne by the contractor but otherwise by the employer.

Clause 39
This clause empowers the engineer to order the contractor to remove and make good improper work or materials which are not in accordance with the contract. This power exists notwithstanding the prior approval of such work by the engineer or his representative. If the contractor does not comply with the engineer's order the employer has the right to have this work done by others and to deduct expenses so incurred from monies due to the contractor.

Clause 48
This clause deals with the issue of completion certificates by the

engineer. Under 48(1) a completion certificate shall be given by the engineer when the whole of the works are substantially completed on a notice given to that effect by the contractor with an undertaking to finish the outstanding work during the period of maintenance; under 48(2) in a similar manner a certificate of completion shall be given for a section for which a separate time for completion is provided in the appendix to the form of tender or for a substantial part of the Works which has been completed to the satisfaction of the engineer and occupied or used by the employer; under 48(3) a certificate of completion can be given for any part of the works which the engineer is of the opinion is substantially complete; the outstanding work in that part is deemed to have been undertaken by the contractor upon issue of such certificate.

Clause 49
This clause which comprises five sub-clauses deals with maintenance and defects. The period of maintenance is defined as the period named in the appendix to the form of tender calculated from the date of completion of the works certified by the engineer. The clause also covers the obligations of the contractor in respect of work to be executed during the period of maintenance, and the cost of execution of such work, remedies for failure and the temporary reinstatement of highways for which the contractor is responsible under the Public Utilities Street Works Act 1950.

Clause 50
This clause empowers the engineer to order the contractor during the period of maintenance to search for any defects that may be observed in the works and to rectify them. In a similar manner to clause 38, if the contractor is liable for such defects he is required to bear the cost of searching and rectification but otherwise such costs are to be borne by the employer.

Clause 51
This clause empowers the engineer to order any variation to any part of the works. A point to note is that such variations should be in the engineer's opinion necessary for the completion of the works.

Clause 52
This clause deals with the manner in which ordered variations are to be valued, the power of the engineer to fix rates, the manner in which daywork is to be executed and the procedure that is to be

followed in respect of claims that are intended to be lodged by the contractor.

Clause 53
This clause deals with vesting of contractor's plant and materials in the employer when these are brought to site.

Clause 54
This clause deals with the vesting in the employer of goods and materials which are not on site. The clause becomes operative only if the contractor lists the the goods and materials in the appendix to the form of tender with a view to obtaining payment for them before they are brought to site.

Clause 61
This clause deals with the issue by the engineer to the employer of the maintenance certificate at the end of the maintenance period. This certificate sets out the date on which the contractor has completed his obligations to construct, complete and maintain the works to the engineer's satisfaction.

Time within which the Works are to be executed

Clause 41
This clause deals with the date for commencement of the works. The date for commencement of the works is the date notified by the engineer to the contractor in writing, which date has to be within a reasonable time after accepting the tender. It is usual for the possible Date for Commencement to be discussed between the engineer and the contractor before the date is notified.

Clause 43
This clause stipulates that the whole of the works or any section thereof shall be completed within the time set out in the appendix to the form of tender (or extended time as may be allowed under clause 44) calculated from the date of commencement.

Clause 44
This clause sets out the procedure for granting extensions of time for the completion of the works.

Other clauses
The other clauses which specifically refer to an entitlement of
extension of time for completion under clause 44 are
   Clause   7: further drawings and instructions
   Clause 12: adverse physical conditions and artificial
      obstructions
   Clause 13(3): delay and extra cost
   Clause 14(2): revision of programme
   Clause 40: suspension of works
   Clause 42: possession of site
   Clause 59B(4): delay and extra cost.

Payment for work executed

Clause 52
This clause deals with the procedure for valuation of ordered
variations. The engineer's powers to fix rates for such variations
are set out in sub-clause (2) and the method of payment for works
carried out under daywork in sub-clause (3).

Clause 56
This clause requires the engineer to ascertain and determine the
value of the work done in accordance with the contract on an
admeasurement basis.

Clause 60
This clause sets out the manner in which monthly statements for
certification and payment should be provided by the contractor.
The procedure for certification; the obligations of the employer,
the contractor and the duties of the engineer in relation to the final
account; together with a series of provisions which include matters
relating to retention monies, interest on overdue payments and
correction and withholding of certificates are also stated in this
clause.

Other clauses
The other clauses which particularly have an effect on the manner
in which payment for work carried out is to be effected may be
stated as follows:
   Clause 55: quantities and correction of errors
   Clause 58: payment in respect of provisional sums and prime
               cost items.
   Clause 59A(5): payments for work carried out on the basis of
                  nominated sub-contractors

Clause 69: tax fluctuations
Clause 70: value added tax
Special Condition: contract price fluctuations.

Default and disputes

Clause 47
This clause lays down the procedure for the imposition of liquidated damages on the contractor for failing to complete a section of or the whole of the works within the time prescribed in the contract. Points to note are: that 'liquidated damages for delay' should represent the employer's genuine and realistic pre-estimate of the damages likely to be suffered by him in the event the works or sections of the works are not completed within the prescribed time and that these damages are not payable as a penalty; secondly, the sums payable as liquidated damages should be stated in the appendix to the form of tender.

Clause 62
This clause empowers the employer on the advice of the engineer to have urgent repairs, which the contractor is unwilling to do, to be attended to by others. If the contractor is liable for such repairs the costs are recoverable from him.

Clause 63
This clause sets out the reasons and the procedure under which a contractor could be expelled from the site. This clause is invoked only as a last resort. The clause empowers the engineer to certify in writing to the employer that one or more of the breaches set out in the clause have been made by the contractor and it is for the employer to expel the contractor from the site. The rights, obligations and liabilities which flow from the contract after expelling a contractor are also set out in this clause.

Clause 66
This clause sets out the procedure under the contract for the settlement of disputes by arbitration. The subject of arbitration and the related procedure is dealt with in Chapter 14.

Other clauses
The other clauses in the conditions of contract which deal directly with default are:

Clause 64: frustration
Clause 65: war clause.

General conditions of contract for work abroad

General conditions of contract of various forms are used in the execution of civil engineering works abroad. The indigenous governmental authorities tend to have a standard form drafted by the legal departments of the country concerned, while some statutory authorities tend to adopt tailor-made conditions to suit their particular needs. All such forms are drafted to incorporate the requirements inherent in civil engineering construction. It follows that the principal clauses provided in them are broadly similar to those adopted by the ICE form with modifications as appropriate to suit local conditions.

There are two standard forms of contract generally used on civil engineering works abroad. These are the ICE Overseas Conditions and the FIDIC Conditions. The Overseas (Civil) Conditions were first published in 1956 as a complementary form of contract to the UK ICE Conditions: this form was prepared by the Association of Consulting Engineers jointly with the Export Group for Constructional Industries and has the approval of the Institution of Civil Engineers (ICE). These Conditions are still in its first edition although several reprints have been made. This document is now largely displaced by the FIDIC Conditions, although it is still the preferred form for use on civil engineering contracts in some Middle East countries.

The FIDIC (Civil) Conditions of Contract were first published in 1957. It is now in its fourth edition, published in 1987. The form is prepared by the International Federation of Consulting Engineers (FIDIC) in consultation with the European Internation-al Contractors (EIC) as the mandatory of Confederation of International Contractors' Asociations (CICA) with participation of Associated General Contractors of America (AGC). FIDIC (Civil) Conditions are widely accepted and used throughout the world. They have been translated into several languages, and the version in English is considered by FIDIC as the official and authentic text for purposes of translation.

Although the FIDIC and Overseas Conditions are adopted, especially by British consulting engineers for works abroad, several projects are carried out today on the basis of discrete forms which follow the pattern of ICE, Overseas and FIDIC documents.

Overseas (Civil) Conditions of Contract

The Overseas (Civil) Conditions of Contract, published in August 1956, closely follow the pattern of the fourth edition of the ICE Conditions which was published a year earlier in 1955. Included with these Conditions are a form of tender with an appendix thereto and a form of agreement. A form of bond is not included. The Conditions are in two parts, Part I being the General Conditions with Part II devoted to conditions of particular application. A set of guidance notes for the preparation of clauses for Part II are included in the published document. Part II (Conditions of Particular Application) is in Appendix 6.14 to this chapter.

Some clauses in Part I – General Conditions are left open to be drafted to suit conditions applicable at the location of the works. These include regulations governing importation of equipment, bribery and corruption, photographs of the works, advertising, classified information, and other matters as appropriate. A clause stipulating the law governing the contract is a mandatory requirement which has to be set out in Part II.

FIDIC Conditions of Contract for works of civil engineering construction

The first edition of the FIDIC (Civil) Conditions of Contract, published in 1957, was based generally on the Overseas (Civil) Conditions of Contract published in 1956, which in turn closely followed the pattern of the fourth edition of the ICE Conditions published in 1955. The second edition of these conditions comprise Parts I and II of the first edition with an additional Part III containing conditions of particular application to dredging and reclamation work. The third edition, published in 1977, was a complete revision, which this time generally was based upon the fifth edition of the ICE Conditions published in 1973.

The current version of the FIDIC (Civil) Conditions of Contract is the fourth edition, published in 1987. This documentation is presented in two parts bound into separate books and placed within one jacket. Part I remains the General Conditions with a Part II into which Part III of the previous edition is incorporated.

Part I of the new edition retains the clause numbering of the third edition and is sensibly similar in content, although almost every clause in it is somewhat modified. Part II of the documentation has been expanded in this edition to include a comprehensive set of guidelines, instructions and worked examples for drafting clauses of particular application relevant to the works. A Form of Tender and Appendix together with a Form of Agreement are included in Part I.

Part I of the fourth edition which comprises the General Conditions are linked with the Conditions of Particular Application contained in Part II by corresponding clauses. Part II must be specially drafted to suit each individual contract. Parts I and II comprise the Conditions that govern the contract.

Comments on some changes in the fourth edition of FIDIC (Civil) Conditions of Contract from its previous edition are set out below.

Clause 1 – Definitions
The number of definitions has been increased. Some terms which are used in the FIDIC Conditions of Contract for Electrical and Mechanical Works have replaced the terms imparting the same meaning used in the third edition of the FIDIC (Civil) Conditions.

Clause 2 – Engineer and engineer's representative
The clause has been expanded and the requirements for the engineer to act impartially has been expressly stated at sub-clause 2.6.

Clause 7.2 – Permanent works designed by contractor
This is a new sub-clause setting out the procedure to be followed by the contractor to obtain approval of the engineer on the part of permanent works expressly required to be designed by him under the contract. The requirement for the contractor to submit operation and maintenance manuals and as constructed drawings is also stipulated in this sub-clause.

Clause 10 – Performance security
The term performance bond is now called performance security. Sub-clause 10.3 imposes an obligation upon the employer to notify the contractor stating the nature of the default prior to making a claim under the performance security.

Sub-clause 14.3 – Cash flow estimation
A new requirement is imposed on the contractor to provide quarterly cash flow estimates of his entitlements under the contract.

Sub-clause 19.1 – Safety, security and protection of the environment
Clause 19 of the previous edition on watching and lighting has been divided into two sub-clauses. 19.1(c) imposes an obligation on the contractor for the protection of the environment: 19.2 deals with the employer's responsibilities for work carried out on the site by his own workmen or through other contractors.

Sub-clause 20.4 – Employer's risks
Sub-clause 20.4 lists a number of risks for which the employer is responsible. These were previously referred to as excepted risks. The previous text has been rearranged and amplified.

Clause 21
The previous clause 21 has been rearranged and expanded into four sub-clauses. Sub-clause 21.1 requires the insurance to be effected by the contractor without limiting his or the employer's obligations under clause 20, for the full replacement cost. Sub-clause 21.2 deals with the scope of cover; a new principle has been introduced in that the risks of both the employer and the contractor have to be covered by a single policy arranged by the contractor in the joint names of both parties.

Clause 23 – Third party insurance
Third party insurance is now to be in the joint names of the employer and the contractor.

Sub-clause 25.1 – Evidence and terms of insurances
The contractor is now required to provide evidence of insurance before starting work on-site.

Clause 34 – Engagement of staff and labour
Clause 34 in the fourth edition on engagement of staff and labour is identical in content to sub-clause 34(1) in the third edition. Sub-clauses 34(2) to (9) and the notes for guidance given in Part II of that document have been replaced in the corresponding part of the fourth edition by a series of comprehensive example sub-clauses. These additional sub-clauses as may be necessary can be incorporated, when Part II of the Conditions of Contract are drafted for a particular contract. Clause 34 of FIDIC Part II is in Appendix 6.14 to this chapter.

Clause 44 – Extension of time
This clause has been expanded and rearranged into three sub-clauses. Sub-clauses 44.3 provides for interim determination of extensions.

Clause 48 – Taking-over certificate
The term certificate of completion used in the third edition is now called the taking-over certificate.

Clause 49 – Defects liability period
The term period of maintenance used in the third edition is now called the defects liability period.

Clause 51 – Variations
Sub-clause 51.1(b) now makes it clear that the engineer is not empowered to instruct the contractor to omit any part of the work if such work is to be carried out by the employer or by another contractor. The new sub-clause 51.1(f) empowers the engineer to change the specified sequence or timing of construction.

Clause 52 – Valuation of variations
This clause has now given express authority for the engineer to fix provisional rates or prices to enable on-account payments to be certified. Sub-clause 52.3 provides for the contract price to be adjusted if the net effect of all variations amount to a variation of 15 per cent of the effective contract price. This percentage, which was reduced from 15 to 10 in the third edition, now reverts back to 15.

Clause 53 – Procedure for claims
The procedure for claims which was covered in sub-clause 52(5) in the third edition is now covered comprehensively under clause 53, which comprises five sub-clauses devoted to this subject. Time limits for notification and substantiation of claims are now imposed and the requirements for the maintenance of contemporary records is explicitly stipulated.

Clause 54 – Contractor's equipment, temporary works and materials
The subject matter covered in clause 53 of the previous edition is now covered under clause 54, as clause 53 is now devoted to claims. Additional sub-clauses are included to also cover conditions of hire relating to contractor's equipment.

Clause 60 – Certificates and payment
A comprehensive clause comprising ten sub-clauses are now set out in respect of certificates and payment in Part I. In the previous edition this clause was left to be drafted for a particular contract in accordance with the guidelines set out at Part II of that document. The present clause in Part I of the fourth edition sets out precise payment procedures with time limits and safeguards for the issue of monthly statements, settlement of the final account and other matters on this subject. In Part II of these Conditions example

clauses which may be necessary to cover certain matters relating to payment are given. These may be used as additional sub-clauses when Part II of the Conditions are drafted for a particular contract. Clause 60 of FIDIC Part II is in Appendix 6.15 to this chapter.

Clause 66 – Release from performance
The word 'frustration' used in the third edition is now substituted by 'release from performance'. Otherwise the content of the clause remains the same.

Clause 67 – Settlement of disputes
Sub-clause 67.2 requires an attempt to be first made for an amicable settlement of a dispute before it could be referred to arbitration.

Clause 69 – Default of employer
Two new sub-clauses 69.4 and 69.5 are added which entitles the contractor to suspend or reduce the rate of work if the employer fails to pay in accordance with the engineer's certificate. The contractor's right to terminate the contract under sub-clause 69.1(a) is not prejudiced by this provision.

Commentary on General Conditions of Contract

The standard forms reviewed above are the ICE Conditions of Contract (fifth edition), the Overseas (Civil) Conditions of Contract (first edition) and the FIDIC (Civil) Conditions of Contract (fourth edition). All these are generally geared to admeasurement contracts where a bill of quantities is included.

For the execution of lump-sum, target cost, cost/reimbursement and other types of contract these standard forms must be modified to suit the particular type of contract. Preferably, such modifications should be set out as special conditions forming a discrete part of the conditions of contract in which should be stipulated any additions, deletions or amendments that are necessary to those contained in the standard form. Such a procedure enables tenderers who are normally familiar with standard forms to readily recognize the changes that have been effected to them.

Other standard forms frequently used on civil engineering and building works are the General Conditions of Government

Contracts for Building and Civil Engineering Works, Form GC/Works/1, Form GC/Works/2 (minor works) and Form C 1001 (building; civil, mechanical and electrical engineering small works), which are used by all central government departments other than the Department of Transport, which mainly uses the ICE Conditions of Contract. The Institution of Civil Engineers has issued two other standard forms for civil engineering work. One is the Conditions of Contract for Ground Investigation, published in 1983, and the other is the Conditions of Contract for Minor Works, which was issued in 1988. The former is for use on ground-investigation contracts in the United Kingdom and are based, where applicable, on the form and policy of the fifth edition of the ICE Conditions of Contract. The ICE Minor Works form is intended for use on works of a simple and straightforward nature where the duration of the contract is generally less than six months and the value does not exceed £100 000.

For building contracts the most commonly used form is the JCT Standard Form of Building Contracts, which is the building equivalent of the ICE civil engineering works form. The JCT form is issued by the Joint Contracts Tribunal, representing the Royal Institute of British Architects, the Royal Institution of Chartered Surveyors, the National Federation of Building Trade Employees, the Association of Consulting Engineers and a few other associations concerned with the construction industry. This has eight versions. These, *inter alia*, include standard forms for use on contracts with quantities, without quantities, to a contractor's design and for those let on a management contract basis.

In respect of mechanical and electrical work the Institution of Electrical Engineers (IEE)/Institution of Mechanical Engineers (IMechE), Model Forms A, B1, B2, B3, C and E are available. These cover Home and Export contracts with and without erection. For international work the FIDIC Conditions of Contract (international) for Electrical and Mechanical Works including erection on-site is often used. For process industries the Institution of Chemical Engineers publish model forms of Conditions of Contract for lump-sum contracts and also for cost/reimbursement contracts.

Many projects involve several engineering disciplines. It follows that often separate contracts in the fields of civil, mechanical and electrical engineering have to be let on the same project. The successful realization of such an undertaking depends, among other things, on ensuring the compatibilty among the forms of contract adopted and the effective coordination of the several contracts involved in the project.

List of Appendices

Appendix 6.1

SHORT DESCRIPTION
OF WORKS:—

All Permanent and Temporary Works in connection with*...

Form of Tender (ICE)

(NOTE: The Appendix forms part of the Tender)

To ....

....

....

GENTLEMEN,

Having examined the Drawings, Conditions of Contract, Specification and Bill of Quantities for the construction of the above-mentioned Works (and the matters set out in the Appendix hereto), we offer to construct and complete the whole of the said Works and maintain the Permanent Works in conformity with the said Drawings, Conditions of Contract, Specification and Bill of Quantities for such sum as may be ascertained in accordance with the said Conditions of Contract.

We undertake to complete and deliver the whole of the Permanent Works comprised in the Contract within the time stated in the Appendix hereto.

If our tender is accepted we will, when required, provide two good and sufficient sureties or obtain the guarantee of a Bank or Insurance Company (to be approved in either case by you) to be jointly and severally bound with us in a sum equal to the percentage of the Tender Total as defined in the said Conditions of Contract for the due performance of the Contract under the terms of a Bond in the form annexed to the Conditions of Contract.

Unless and until a formal Agreement is prepared and executed this Tender, together with your written acceptance thereof, shall constitute a binding Contract between us.

We understand that you are not bound to accept the lowest or any tender you may receive.

We are, Gentlemen,

Yours faithfully,

Signature...

Address...

....

Date ....

* Complete as appropriate

Appendix 6.2

FORM OF TENDER (APPENDIX) (I.C.E.)

APPENDIX

NOTE: Relevant Clause numbers are shown in brackets following the description

Amount of Bond (if any)   (10)   ....% of Tender Total

Minimum Amount of Insurance   (23 (2))   £.............

Time for Completion   (43)   Liquidated Damages for Delay   (47)

Column 1
(see Clause 47 (1)

For the Whole of the Works————(a)Weeks   £............(b)   per Day/Week(c)

For the following Sections   Column 2   Column 3
                                        (see Clause 47 (2))

Section(d) ....   £............   £............

............Weeks   Per Day/Week(c)   per Day/Week(c)

Section(d) ....   £............   £............

............Weeks   per Day/Week(c)   per Day/Week(c)

Section(d) ....   £............   £............

............Weeks   per Day/Week(c)   per Day/Week(c)

Section(d) ....   £............   £............

............Weeks   per Day/Week(c)   per Day/Week(c)

Period of Maintenance   (49 (1))   ....Weeks

Vesting of Materials not on Site   (54 (1) and 60 (1))(e)

1............   4............

2............   5............

3............   6............

Standard Method of Measurement adopted in preparation of Bills of Quantities (57)(f)............

....

Percentage for adjustment of P.C. Sums   (59A (2)(b) and (5) (c))   ....%

Percentage of the Value of Goods and Materials to be included in Interim Certificates   (60 (2)(b))   ....%

Minimum Amount of Interim Certificates   (60 (2))   £............

(a) To be completed in every case (by Contractor if not already stipulated).
(b) To be completed by Engineer in every case.
(c) Delete which not required.
(d) To be completed if required, with brief description.
(e) (If used) materials to which clauses apply are to be filled in by Engineer prior to inviting tenders.
(f) Insert here any amendment or modification adopted if different from that stated in Clause 57.

Appendix 6.3

Form of Tender (FIDIC)

NAME OF CONTRACT: *_____

TO: *_____

_____

_____

Gentlemen,

1. Having examined the Conditions of Contract, Specification, Drawings, and Bill of Quantities and Addenda Nos _____ for the execution of the above-named Works, we the undersigned, offer to execute and complete such Works and remedy any defects therein in conformity with the Conditions of Contract, Specification, Drawings, Bill of Quantities and Addenda for the sum of

    _____

    _____

    (_____)
    or such other sums as may be ascertained in accordance with the said Conditions.

2. We acknowledge that the Appendix forms part of our Tender.

3. We undertake, if our Tender is accepted, to commence the works as soon as is reasonably possible after the receipt of the Engineer's notice to commence, and to complete the whole of the Works comprised in the Contract within the time stated in the Appendix to Tender.

4. We agree to abide by this Tender for the period of *_____ days from the date fixed for receiving the same and it shall remain binding upon us and may be accepted at any time before the expiration of that period.

5. Unless and until a formal Agreement is prepared and executed this Tender, together with your written acceptance thereof, shall constitute a binding contract between us.

6. We understand that you are not bound to accept the lowest or any tender you may receive.

    Dated this _____ day of _____ 19 _____

    Signature _____ in the capacity of _____

    duly authorised to sign tenders for and on behalf of _____

    _____

    (IN BLOCK CAPITALS)

    Address _____

    _____

    Witness _____
    Address _____

    _____

    Occupation _____

    (Note: All details marked * shall be inserted before issue of Tender documents.)

Appendix 6.4

Appendix (to Form of Tender – FIDIC)

	Sub-Clause	
Amount of security (if any) _____	**10.1**	_____ per cent of the Contract Price
Minimum amount of third party insurance	**23.2**	_____ per occurrence, with the number of occurrences unlimited
Time for issue of notice to commence	**41.1**	_____ days
Time for Completion _____	**43.1**	_____ days
Amount of liquidated damages _____	**47.1**	_____ per day
Limit of liquidated damages _____	**47.1**	_____
Defects Liability Period _____	**49.1**	_____ days
Percentage for adjustment of Provisional Sums	**59.4(c)**	_____ per cent
Percentage of invoice value of listed materials	**60.1(c)**	_____ per cent
Percentage of Retention _____	**60.2**	_____ per cent
Limit of Retention Money _____	**60.2**	_____
Minimum Amount of Interim Certificates __	**60.2**	_____
Rate of interest upon unpaid sums _____	**60.10**	_____ per cent

Initials of Signatory of Tender _____

(Notes: All details in the list above, other than percentage figure against Sub-Clause 59.4, shall be inserted before issue of Tender documents. Where a number of days is to be inserted, it is desirable, for consistency with the Conditions, that the number should be a multiple of seven.

Additional entries are necessary where provision is included in the Contract for:

(a) completion of Sections (Sub-Clauses 43.1 and 48.2(a))
(b) liquidated damages for Sections (Sub-Clause 47.1)
(c) a bonus (Sub-Clause 47.3 — Part II)
(d) payment for materials on Site (Sub-Clause 60.1(c))
(e) payment in foreign currencies (Clause 60 — Part II)
(f) an advance payment (Clause 60 — Part II)
(g) adjustments to the Contract Price on account of Specified Materials (Sub-Clause 70.1 — Part II)
(h) rates of exchange (Sub-Clause 72.2 — Part II))

Appendix 6.5

**SHORT DESCRIPTION
OF WORKS.**

Form of Tender (OVERSEAS)

(NOTES :—The Appendix forms part of the Tender.

Tenderers are required to fill up all the blank spaces in this Tender Form and Appendix.)

To : ....

GENTLEMEN,

Having examined the Drawings, Conditions of Contract, Specification and Bill of Quantities for the construction of the above-named Works, We, the undersigned, offer to construct complete and maintain the whole of the said Works in conformity with the said Drawings, Conditions of Contract, Specification and Bill of Quantities for the sum of...

.... (£.........................)

or such other sum as may be ascertained in accordance with the said Conditions.

2. We undertake if our Tender is accepted to commence the Works within...................days of receipt of the Engineer's order to commence, and to complete and deliver the whole of the Works comprised in the Contract within...................days calculated from the last day of the aforesaid period in which the Works are to be commenced.

3. If our tender is accepted we will, if required, provide two good and sufficient sureties or obtain the guarantee of a Bank or Insurance Company (to be approved in either case by you) to be jointly and severally bound with us in a sum not exceeding 10 per cent. of the above-named sum for the due performance of the Contract under the terms of a Bond to be approved by you.

4. We agree to abide by this Tender for the period of...................days from the date fixed for receiving the same and it shall remain binding upon us and may be accepted at any time before the expiration of that period.

5. Unless and until a formal Agreement is prepared and executed this Tender, together with your written acceptance thereof, shall constitute a binding Contract between us.

6. We understand that you are not bound to accept the lowest or any tender you may receive.

Appendix 6.6

Appendix (to Form of Tender – Overseas)

CLAUSE

Amount of Bond or Guarantee (if any) ... ... 10 £.......................................

Minimum Amount of Third Party Insurance ... ... 23 (2) £.......................................

Period for commencement, from Engineer's order to
commence ... ... 41 .... days

Time for Completion ... ... 43 .... days

Amount of Liquidated Damages ... ... 47 (1) £....................per day

Period of Maintenance ... ... 49 .... days

Percentage for Adjustment of P.C. Sums ... ... 58 (2) ....per cent.

Percentage of Retention ... ... 60 ( ) ....per cent.

Limit of Retention Money ... ... 60 ( ) £..............................

Minimum Amount of Interim Certificates ... ... 60 ( ) £..............................

Time within which payment to be made after Certificate 60 ( ) ....days

Appointor of Arbitrator ... ... 66 ....

Dated this..day of..., 19 ......

Signature... in the capacity of...........................

duly authorised to sign tenders for and on behalf of ....

....

(IN BLOCK CAPITALS)

Witness.. Address...

Address.. ....

....

Occupation ....

Appendix 6.7

Form of Agreement (I.C.E.)

THIS AGREEMENT made the .... day of ....
19............ BETWEEN...
of .... in the
County of .... (hereinafter called " the Employer ") of the one part and
.... of ....
in the County of ....
.... (hereinafter called " the Contractor ") of the other part
WHEREAS the Employer is desirous that certain Works should be constructed, viz. the Permanent
and Temporary Works in connection with...
....and has accepted a Tender by the Contractor for
the construction and completion of such Works and maintenance of the Permanent Works
NOW THIS AGREEMENT WITNESSETH as follows:—

1. In this Agreement words and expressions shall have the same meanings as are respectively
assigned to them in the Conditions of Contract hereinafter referred to.

2. The following documents shall be deemed to form and be read and construed as part of
this Agreement, viz:—

    (a) The said Tender.
    (b) The Drawings.
    (c) The Conditions of Contract.
    (d) The Specification.
    (e) The Priced Bill of Quantities.

3. In consideration of the payments to be made by the Employer to the Contractor as
hereinafter mentioned the Contractor hereby covenants with the Employer to construct and
complete the Works and maintain the Permanent Works in conformity in all respects with the
provisions of the Contract.

4. The Employer hereby covenants to pay to the Contractor in consideration of the construction and completion of the Works and maintenance of the Permanent Works the Contract
Price at the times and in the manner prescribed by the Contract.

IN WITNESS whereof the parties hereto have caused their respective Common Seals to be
hereunto affixed (or have hereunto set their respective hands and seals) the day and year first above
written

The Common Seal of...
.... Limited
was hereunto affixed in the presence of:—

or

SIGNED SEALED AND DELIVERED by the
said ....
....

in the presence of:—

....
....

Appendix 6.8

Form of Agreement (FIDIC)

This Agreement made the _____ day of _____ 19 ____

Between _____

of _____

_____ (hereinafter called "the Employer) of the one part and

_____ of _____

(hereinafter called the "Contractor") of the other part

Whereas the Employer is desirous that certain Works should be executed by the contractor, viz _____

and has accepted a Tender by Contractor for the execution and completion of such Works and the remedying of any defects therein

Now this agreement witnesseth as follows:

1. In this Agreement words and expressions shall have the same meanings as are respectively assigned to them in the Conditions of Contract hereinafter referred to.

2. The following documents shall be deemed to form and be read and construed as part of this Agreement, viz:-

   (a) The Letter of Acceptance;
   (b) The said Tender;
   (c) The Conditions of Contract (Parts I and II);
   (d) The Specification;
   (e) The Drawings; and
   (f) The Bill of Quantities.

3. In consideration of the payments to be made by the Employer to the Contractor as hereinafter mentioned the Contractor hereby covenants with the Employer to execute and complete the Works and remedy any defects therein in conformity in all respects with the provisions of the Contract.

4. The Employer hereby covenants to pay the Contractor in consideration of the execution and completion of the Works and the remedying of defects therein the Contract Price or such other sum as may become payable under the provisions of the Contract at the times and in the manner prescribed by the Contract.

In Witness whereof the parties hereto have caused this Agreement to be executed the day and year first before written in accordance with their respective laws.

The Common Seal of _____

_____

was hereunto affixed in the presence of:-

or

Signed Sealed and Delivered by the

said _____

in the presence of:

Binding Signature of Employer _____

Binding Signature of Contractor _____

Appendix 6.9

Form of Agreement (OVERSEAS)

THIS AGREEMENT made the .... day of ....…........

19......... BETWEEN..

of .... in the

County of ....(hereinafter called " the Employer ") of the one part and.........

.... of ....

in the County of ....(hereinafter called " the Contractor ") of the other part

WHEREAS the Employer is desirous that certain Works should be constructed, viz.....................

....and has accepted a

Tender by the Contractor for the construction completion and maintenance of such Works NOW

THIS AGREEMENT WITNESSETH as follows :—

1.   In this Agreement words and expressions shall have the same meanings as are respectively assigned to them in the Conditions of Contract hereinafter referred to.

2.   The following documents shall be deemed to form and be read and construed as part of this Agreement, viz.:—

     (a) The said Tender.

     (b) The Drawings.

     (c) The Conditions of Contract (Parts I and II).

     (d) The Specification.

     (e) The Bill of Quantities.

     (f) The Schedule of Rates and Prices (if any).

3.   In consideration of the payments to be made by the Employer to the Contractor as hereinafter mentioned the Contractor hereby covenants with the Employer to construct complete and maintain the Works in conformity in all respects with the provisions of the Contract.

4.   The Employer hereby covenants to pay the Contractor in consideration of the construction completion and maintenance of the Works the Contract Price at the times and in the manner prescribed by the Contract.

IN WITNESS whereof the parties hereto have caused their respective Common Seals to be hereunto affixed (or have hereunto set their respective hands and seals) the day and year first above written

The Common Seal of....................................

.... Limited

was hereunto affixed in the presence of :—

        or

SIGNED SEALED AND DELIVERED by the

said..

....

in the presence of :—

Appendix 6.10

Form of Bond (ICE)

BY THIS BOND [1]We ....

of .... in the

County of .... [2]We   Limited

whose registered office is at .... in the

County of .... [3]We

and .... carrying on business in partnership under

the name or style of ....

at .... in the

County of .... (hereinafter called " the Contractor ") [4]and

.... of ....

in the County of .... and ....

of .... in the County of ....

.... [5]and   Limited

whose registered office is at .... in the

County of .... (hereinafter called " the [4]Sureties/Surety ") are held and firmly

bound unto .... (hereinafter

called " the Employer ") in the sum of .... pounds

(£ ....) for the payment of which sum the Contractor and the [4]Sureties/Surety bind

themselves their successors and assigns jointly and severally by these presents.

Sealed with our respective seals and dated this .... day of ....

19 ....

[1] Is appropriate
to an individual,
2 to a Limited
Company and 3 to
a Firm. Strike
out whichever two
are inappropriate.

[4] Is appropriate
where there are
two individual
Sureties, 5 where
the Surety is a
Bank or Insurance
Company. Strike
out whichever is
inappropriate.

WHEREAS the Contractor by an Agreement made between the Employer of the one part and the Contractor of the other part has entered into a Contract (hereinafter called "the said Contract ") for the construction and completion of the Works and maintenance of the Permanent Works as therein mentioned in conformity with the provisions of the said Contract.

NOW THE CONDITIONS of the above-written Bond are such that if:—
  (a) the Contractor shall subject to Condition (c) hereof duly perform and observe all the terms provisions conditions and stipulations of the said Contract on the Contractor's part to be performed and observed according to the true purport intent and meaning thereof or if
  (b) on default by the Contractor the Sureties/Surety shall satisfy and discharge the damages sustained by the Employer thereby up to the amount of the above-written Bond or if
  (c) the Engineer named in Clause 1 of the said Contract shall pursuant to the provisions of Clause 61 thereof issue a Maintenance Certificate then upon the date stated therein (hereinafter called " the Relevant Date ")
this obligation shall be null and void but otherwise shall remain in full force and effect but no alteration in the terms of the said Contract made by agreement between the Employer and the Contractor or in the extent or nature of the Works to be constructed completed and maintained thereunder and no allowance of time by the Employer or the Engineer under the said Contract nor any forbearance or forgiveness in or in respect of any matter or thing concerning the said Contract on the part of the Employer or the said Engineer shall in any way release the Sureties/Surety from any liability under the above-written Bond.

PROVIDED ALWAYS that if any dispute or difference shall arise between the Employer and the Contractor concerning the Relevant Date or otherwise as to the withholding of the Maintenance Certificate then for the purposes of this Bond only and without prejudice to the resolution or

Appendix 6.10 continued

determination pursuant to the provisions of the said Contract of any dispute or difference whatsoever between the Employer and Contractor before the Relevant Date shall be such as may be:—

    (a)   agreed in writing between the Employer and the Contractor or

    (b)   if either the Employer or the Contractor shall be aggrieved at the date stated in the said Maintenance Certificate or otherwise as to the issue or withholding of the said Maintenance Certificate the party so aggrieved shall forthwith by notice in writing to the other refer any such dispute or difference to the arbitration of a person to be agreed upon between the parties or (if the parties fail to appoint an arbitrator within one calendar month of the service of the notice as aforesaid) a person to be appointed on the application of either party by the President for the time being of the Institution of Civil Engineers and such arbitrator shall forthwith and with all due expedition enter upon the reference and make an award thereon which award shall be final and conclusive to determine the Relevant Date for the purposes of this Bond. If the arbitrator declines the appointment or after appointment is removed by order of a competent court or is incapable of acting or dies and the parties do not within one calendar month of the vacancy arising fill the vacancy then the President for the time being of the Institution of Civil Engineers may on the application of either party appoint an arbitrator to fill the vacancy. In any case where the President for the time being of the Institution of Civil Engineers is not able to exercise the aforesaid functions conferred upon him the said functions may be exercised on his behalf by a Vice-President for the time being of the said Institution.

        Signed Sealed and Delivered by the said ⎫
          in the presence of:—   ⎭

        The Common Seal of   ⎫
                       LIMITED ⎬
        was hereunto affixed in the presence of:— ⎭

        (Similar forms of Attestation Clause for the Sureties or Surety)

Appendix 6.11

EXAMPLE SURETY BOND FOR PERFORMANCE (FIDIC)

Know all Men by these Presents that (name and address of Contractor)

_____

_____

as Principal (hereinafter called "the Contractor") and (name, legal title and address of Surety) _____

_____

as Surety (herein after called "the Surety"), are held and firmly bound unto (name and address of Employer) _____

_____ *as Obligee (hereinafter called "the Employer") in the amount of* _____ *for the payment of which sum, well and truly to be made, the Contractor and the Surety bind themselves, their successors and assigns, jointly and severally, firmly by these presents.*

Whereas the Contractor has entered into a written contract agreement with the Employer dated the _____ *day of* _____ *19* _____

for (name of Works) _____
in accordance with the plans and specifications and amendments thereto, to the extent herein provided for, are by reference made part hereof and are hereinafter referred to as the Contract.

Appendix 6.11 continued

Now, therefore, the Condition of this Obligation is such that, if the Contractor shall promptly and faithfully perform the said Contract (including any amendments thereto) then this obligation shall be null and void; otherwise it shall remain in full force and effect.

Whenever Contractor shall be, and declared by Employer to be, in default under the Contract, the Employer having performed the Employer's obligations thereunder, the Surety may promptly remedy the default, or shall promptly:-

(1) Complete the Contract in accordance with its terms and conditions; or

(2) Obtain a bid or bids for submission to the Employer for completing the Contract in accordance with its terms and conditions, and upon determination by Employer and Surety of the lowest responsible bidder, arrange for a contract between such bidder and Employer and make available as work progresses (even though there should be a default or a succession of defaults under the contract or contracts of completion arranged under this paragraph) sufficient funds to pay the cost of completion less the balance of the Contract Value; but not exceeding, including other costs and damages for which the Surety may be liable hereunder, the amount set forth in the first paragraph hereof. The term "balance of the Contract Value", as used in this paragraph, shall mean the total amount payable by Employer to Contractor under the Contract, less the amount properly paid by Employer to Contractor; or

(3) Pay the Employer the amount required by Employer to complete the Contract in accordance with its terms and conditions any amount up to a total not exceeding the amount of this Bond.

The Surety shall not be liable for a greater sum than the specified penalty of this Bond.

Any suit under this Bond must be instituted before the issue of the Defects Liability Certificate.

No right of action shall accrue on this Bond to or for the use of any person or corporation other than the Employer named herein or the heirs, executors, administrators or successors of the Employer.

Signed on _____   *Signed on* _____

on behalf of _____   *on behalf of* _____

by _____   *by* _____

in the capacity of _____   *in the capacity of* _____

in the presence of _____   *in the presence of* _____

Appendix 6.12

EXAMPLE PERFORMANCE GUARANTEE (FIDIC)

By this guarantee We, _____

whose registered office is at _____

(hereinafter called "the Contractor") and _____

whose registered office is at _____

(hereinafter called "the Guarantor") are held and firmly bound unto

_____ *(hereinafter called "the Employer")*

in the sum of _____ *for the payment of which sum*

the Contractor and the Guarantor bind themselves, their successors and assigns jointly and severally by these presents.

Whereas the Contractor by an Agreement made between the Employer of the one part and the Contractor of the other part has entered into a Contract (hereinafter called "the said Contract") to execute and, complete certain Works and remedy any defects therein as therein mentioned in conformity with the provisions of the said Contract.

Now the Condition of the above-written Guarantee is such that if the Contractor shall duly perform and observe all the terms provisions conditions and stipulations of the said Contract on the Contractor's part to be performed and observed according to the true purport intent and meaning thereof or if on default by the Contract the Guarantor shall satisfy and discharge the damages sustained by the Employer thereby up to the amount of the above-written Guarantee then this obligation shall be null and void but otherwise shall be and remain in full force and effect but no alteration in terms of the said Contract or in the extent or nature of the Works to be executed, completed and defects therein remedied thereunder and no allowance of time by the Employer or the Engineer under the said Contract nor any forbearance or forgiveness in or in respect of any matter or thing concerning the said Contract on the part of the Employer or the said Engineer shall in any way release the Guarantor from any liability under the above-written Guarantee. Provided always that the above obligation of Guarantor to satisfy and discharge the damages sustained by the Employer shall arise only

(a) on written notice from both the Employer and the Contractor that the Employer and the Contractor have mutually agreed that the amount of damages concerned is payable to the Employer or

(b) on receipt by the Guarantor of a legally certified copy of an award issued in arbitration proceeding carried out in conformity with the terms of the said Contract that the amount of the damages is payable to the Employer.

Signed on _____   *Signed on* _____

on behalf of _____   *on behalf of* _____

by _____   *by* _____

in the capacity of _____   *in the capacity of* _____

in the presence of _____   *in the presence of* _____

Appendix 6.13

Conditions of Contract

PART II—CONDITIONS OF PARTICULAR APPLICATION (OVERSEAS)

The following notes are intended as a guide in the preparation of clauses (some of which are referred to in Part I) which will vary as necessary to take account of the circumstances and locality of the Works.   These variable clauses should cover such of the undermentioned matters as are applicable.

Clause 1—Definitions

The names of the " Employer " and " Engineer " and the Calendar (e.g. Gregorian Calendar) to be used for the Contract and any further definitions required.

Clause 34—Labour

Engagement of labour; permits for imported labour, transport, accommodation; feeding; drinking water; control; camps; health; epidemics; non-supply of arms and liquor; religious customs and festivals; observance of days of rest; prevention of riotous or unlawful behaviour; hours and conditions; rates of pay; compliance with labour legislation.

Clause 49—Temporary Reinstatement

In appropriate cases, where the permanent reinstatement is not being carried out by the Contractor, an additional sub-clause should be added to Clause 49 to cover making good all subsidence etc. in the temporary reinstatement of any highway broken into for the purposes of the execution of the Works and the liability for damage and injury resulting therefrom up to the end of the Period of Maintenance or until possession of the site has been taken for the purpose of carrying out permanent reinstatement (whichever is the earlier).

Clause 53—Property in Materials and Plant

Vesting of Constructional Plant (other than a registered British Ship), Temporary Works and Materials in the Employer for the purposes of the Contract, re-vesting in Contractor on removal with consent of Engineer or on completion or determination of contract; refund of outstanding advances (if any) on removal; liability for damage to Plant; right of re-export of Plant; hire of Plant; sale or disposal of Plant.

Clause 60—Certificates and Payment

Advances on Plant and Materials where made; conditions covering such advances and their repayment; monthly claims for work executed and certificates of Engineer as to amount due to Contractor for permanent work executed in the month and for temporary works included in the Bill of Quantities and also, if there are no advances for materials and plant amounts as certified by the Engineer for any materials for permanent work on the site; times of payment of retention monies; correction and withholding of certificates; currency for payments; place of payment; rate of exchange; times of payment.

Clause 69—Variation of Price (Labour and Materials)

In appropriate cases this clause should cover such matters as :—

Adjustment of contract price by reason of alteration in rates of wages and allowances payable to labour and local staff; change in conditions of employment of labour and local staff; change in cost of materials for permanent or temporary works, or in consumable stores, fuel and power; variation in freight and insurance rates; the operation of any law, statute, etc.

Clause 70—Duties, Dues, etc.

Payment of or relief from Customs or other import duties (to be set out in specific terms), harbour, wharfage, landing, pilotage and any other charges or dues.

Clause 71—Taxation

Payment of or exemption from local income or other taxes both as regards the Contractor and his staff.

Clause 72—Law Governing Contract

This clause should state the country to the law of which the contract is subject and in accordance with which it will be construed.

N.B. This would normally be put at the end of Part II.

Clause 73, etc.—Miscellaneous

In certain cases it may be desirable to insert clauses to cover such matters as :—

(a) regulations governing importation and use of explosives for blasting;

(b) bribery and corruption;

(c) photographs of the Works and advertising;

(d) undertakings regarding non-disclosure of secret information;

(e) any other matters special to the contract.

Appendix 6.14

FIDIC Part II Conditions of Particular Application

Clause 34

It will generally be necessary to add a number of Sub-Clauses, to take account of the circumstances and locality of the Works, covering such matters as: permits and registration of expatriate employees; repatriation to place of recruitment; provision of temporary housing for employees; requirements in respect of accommodation for staff of Employer and Engineer; standards of accommodation to be provided; provision of access roads, hospital, school, power, water, drainage, fire services, refuse collection, communal buildings, shops, telephones; hours and conditions of working; rates of pay; compliance with labour legislation; maintenance of records of safety and health.

EXAMPLE SUB-CLAUSES (to be numbered, as appropriate)

Rates of Wages and Conditions of Labour	34.	*The Contractor shall pay rates of wages and observe conditions of labour not less favourable than those established for the trade or industry where the work is carried out. In the absence of any rates of wages or conditions of labour so established, the Contractor shall pay rates of wages and observe conditions of labour which are not less favourable than the general level of wages and conditions observed by other employers whose general circumstances in the trade or industry in which the Contractor is engaged are similar.*
Employment of Persons in the Service of Others	34.	*The Contractor shall not recruit or attempt to recruit his staff and labour from amongst persons in the service of the Employer or the Engineer.*
Repatriation of Labour	34.	*The Contractor shall be responsible for the return to the place where they were recruited or to their domicile of all such persons as he recruited and employed for the purposes of or in connection with the Contract and shall maintain such persons as are to be so returned in a suitable manner until they shall have left the Site or, in the case of persons who are not nationals of and have been recruited outside (insert name of country) shall have left (insert name of country).*
Housing for Labour	34.	*Save insofar as the Contract otherwise provides, the Contractor shall provide and maintain such accommodation and amenities as he may consider necessary for all his staff and labour, employed for the purposes of or in connection with the Contract, including all fencing, water supply (both for drinking and other purposes), electricity supply, sanitation, cookhouses, fire prevention and fire-fighting equipment, air conditioning, cookers, refrigerators, furniture and other requirements in connection with such accommodation or amenities. On completion of the Contract, unless otherwise agreed with the Employer, the temporary camps/housing provided by the Contractor shall be removed and the site reinstated to its original condition, all to the approval of the Engineer.*
Accident Prevention Officer; Accidents	34.	*The Contractor shall have on his staff at the Site an officer dealing only with questions regarding the safety and protection against accidents of all staff and labour. This officer shall be qualified for this work and shall have the authority to issue instructions and shall take protective measures to prevent accidents.*
Health and Safety	34.	*Due precautions shall be taken by the Contractor, and at his own cost, to ensure the safety of his staff and labour and, in collaboration with and to the requirements of the local health authorities, to ensure that medical staff, first aid equipment and stores, sick bay and suitable ambulance service are available at the camps, housing and on the Site at all times throughout the period of the Contract and that suitable arrangements are made for the prevention of epidemics and for all necessary welfare and hygiene requirements.*
Measures against Insect and Pest Nuisance	34.	*The Contractor shall at all times take the necessary precautions to protect all staff and labour employed on the site from insect nuisance, rats and other pests and reduce the dangers to health and the general nuisance occasioned by the same. The Contractor shall provide his staff and labour with suitable prophylactics for the prevention of malaria and take steps to prevent the formation of stagnant pools of water. He shall comply with all the regulations of the local health authorities in these respects and shall in particular arrange to spray thoroughly with approved insecticide all buildings erected on the Site. Such treatment shall be carried out at least once a year or as instructed by the Engineer. The Contractor shall warn his staff and labour of the dangers of bilharzia and wild animals.*
Epidemics	34.	*In the event of any outbreak of illness of an epidemic nature, the contractor shall comply with and carry out such regulations, orders and requirements as may be made by the Government, or the local medical or sanitary authorities, for the purpose of dealing with and overcoming the same.*

Appendix 6.14 continued

Burial of the Dead	**34.**	The Contractor shall make all necessary arrangements for the transport, to any place as required for burial, of any of his expatriate employees or members of their families who may die in (insert name of country). The Contractor shall also be responsible, to the extent required by the local regulations, for making any arrangements with regard to burial of any of his local employees who may die while engaged upon the Works.
Supply of Foodstuffs	**34.**	The Contractor shall arrange for the provision of a sufficient supply of suitable food at reasonable prices for all his staff and labour, or his Subcontractors, for the purposes of or in connection with the Contract.
Supply of Water	**34.**	The Contractor shall, so far as is reasonably practicable, having regard to local conditions, provide on the Site an adequate supply of drinking and other water for the use of his staff and labour.
Alcoholic Liquor or Drugs	**34.**	The contractor shall not, otherwise than in accordance with the Statutes, Ordinances and Government Regulations or Orders for the time being in force, import, sell, give, barter or otherwise dispose of any alcoholic liquor or drugs, or permit or suffer any such importation, sale, gift, barter or disposal by his Subcontractors, agents, staff or labour.
Arms and Ammunition	**34.**	The Contractor shall not give, barter or otherwise dispose of to any person or persons, any arms or ammunition of any kind or permit or suffer the same as aforesaid.
Festivals and Religious Customs	**34.**	The Contractor shall in all dealings with his staff and labour have due regard to all recognised festivals, days of rest and religious or other customs.
Disorderly Conduct	**34.**	The Contractor shall at all times take all reasonable precautions to prevent any unlawful, riotous or disorderly conduct by or amongst his staff and labour and for the preservation of peace and protection of persons and property in the neighbourhood of the Works against the same.

Appendix 6.15

FIDIC Part II Conditions of Particular Application

Clause 60

Additional Sub-Clauses may be necessary to cover certain other matters relating to payments.

Where payments are to be made in various currencies in predetermined proportions and calculated at fixed rates of exchange the following 3 Sub-Clauses, which should be taken together, may be added:

EXAMPLE SUB-CLAUSES (to be numbered, as appropriate)

Currency of Account and Rates of Exchange	**60.**	The currency of account shall be the (insert name of currency) and for the purposes of the Contract conversion between (insert name of currency) and other currencies stated in the Appendix to Tender shall be made in accordance with the Table of Exchange Rates in the Appendix to Tender. Conversion between the currencies stated in such Table other than the (insert name of currency) shall be made at rates of exchange determined by use of the relative rates of exchange between such currencies and the (insert name of currency) set out therein.
Payments to Contractor	**60.**	All payments to the Contractor by the Employer shall be made
		(a) in the case of payment(s) under Sub-Clause(s) 70.2 and (insert number of any other applicable Clause), in (insert name of currency/ies);
		(b) in the case of payments for certain provisional sum items excluded from the Appendix to Tender, in the currencies and proportions applicable to these items at the time when the Engineer gives instructions for the work covered by these items to be carried out;

Appendix 6.15 continued

(c) in any other case, including Increase or Decrease of Costs under Sub-Clause 70.1, in the currencies and proportions stated in the Appendix to Tender as applicable to such payment provided that the proportions of currencies stated in the Appendix to Tender may from time to time upon the application of either party be varied as may be agreed.

Payments to Employer **60.** *All payments to the Employer by the Contractor including payments made by way of deduction or set-off shall be made*

(a) in the case of credit(s) under Sub-Clause(s) 70.2 and (insert number of any other applicable Clause) in (insert name of currency/ies);

(b) in the case of liquidated damages under Clause 47, in (insert name of currency/ies);

(c) in the case of reimbursement of any sum previously expended by the Employer, in the currency in which the sum was expended by the Employer;

(d) in any other case, in such currency as may be agreed.

If the part payable in a particular currency of any sum payable to the Contractor is wholly or partly insufficient to satisfy by way of deduction or set-off a payment due to the Employer in that currency, in accordance with the provisions of this Sub-Clause, then the Employer may if he so desires make such deduction or set-off wholly or partly as the case may be from the balance of such sum payable in other currencies.

Where all payments are to be made in one currency the following Sub-Clause may be added:

EXAMPLE SUB-CLAUSE (to be numbered, as appropriate)

Currency of Account and Payments **60.** *The currency of account shall be the (insert name of currency) and all payments made in accordance with the Contract shall be in (insert name of currency). Such (insert name of currency), other than for local costs, shall be fully convertible. The percentage of such payments attributed to local costs shall be as stated in the Appendix to Tender.*

Where place of payment is to be defined the following Sub-Clause may be added:

EXAMPLE SUB-CLAUSE (to be numbered, as appropriate)

Place of Payment **60.** *Payments to the Contractor by the Employer shall be made into a bank account nominated by the Contractor in the country of the currency of payment. Where payment is to be made in more than one currency separate bank accounts shall be nominated by the Contractor in the country of each currency and payments shall be made by the Employer accordingly.*

Where provision is to be included for an advance payment the following Sub-Clause may be added:

EXAMPLE SUB-CLAUSE (to be numbered, as appropriate)

Advance Payment **60.** *An advance payment of the amount stated in the Appendix to Tender shall, following the presentation by the Contractor to the Employer of an approved performance security in accordance with Sub-Clause 10.1 and a Guarantee in terms approved by the Employer for the full value of the advance payment, be certified by the Engineer for payment to the Contractor. Such Guarantee shall be progressively reduced by the amount repaid by the Contractor as indicated in interim certificates of the Engineer issued in accordance with this Clause. The advance payment shall not be subject to retention. The advance payment shall be repaid by way of reduction in interim certificates commencing with the next certificate issued after the total certified value of the Permanent Works and any other items in the Bill of Quantities (excluding the deduction of retention) exceeds (insert figure) per cent of the sum stated in the Letter of Acceptance. The amount of the reduction in each interim certificate shall be one (insert fraction) of the difference between the total value of the Permanent Works and any other items in the Bill of Quantities (excluding the deduction of retention) due for certification in such interim certificate and the said value in the last preceding interim certificate until the advance payment has been repaid in full. Provided that upon the issue of a Taking-Over Certificate for the whole of the Works or upon the happening of any of the events specified in Sub-Clause 63.1 or termination under Clauses 65, 66 or 69, the whole of the balance then outstanding shall immediately become due and payable by the Contractor to the Employer.*

Chapter 7

Specifications

The specification is that part of the contract documents which sets out the scope of the works and a detailed description of the nature and quality of the materials and workmanship to be complied with in the execution of the works. When the specification is read in conjunction with the drawings and the bill of quantities the contractor should have all the information that he needs for the proper execution of the works.

The specification also sets forth general information on the project, the facilities that have to be provided at site for the engineer and other contractors, information as to the availability or otherwise of electricity, water and other services required for the execution of the works, the means of access to the site and other special responsibilities the contractor has to bear which are not specifically included in the conditions of contract.

The detailed manner in which programmes and proposed method statements are to be furnished by the contractor is often elaborated in the specification as an amplification of the corresponding clauses of the conditions of contract.

In drafting the specification, due care, wisdom and experience are of great importance in providing clarity of intention. Equally important is the proper implementation of the specification. Failure in these two aspects can be unnecessarily expensive to the employer, who desires a quality end product. The specification should avoid repetition, contradiction or conflict with the other elements of the contract documents and should not include matter that should rightly be stipulated in the instructions to tenderers, conditions of contract, method of measurement or the preamble to the bill of quantities.

In contracts for civil engineering works it is normal practice to describe the nature and quality of the materials and workmanship fully in the specification and keep the description of items in the bill of quantities as brief as possible.

The specification controls the quality of work, and on its soundness the character of the finished work depends to a large degree.

Compilation of the specification

The information for drafting a specification flows from many sources. The particular project for which the specification is drawn up should always be borne in mind and only those clauses which are relevant to the project should be included. The principal sources of information which are commonly used are as follows.

Previous specifications

In most cases specifications adopted for past jobs of a similar type are used as a precedent for the preparation of a new specification.

Tender drawings

The draft tender drawings are frequently prepared ahead of the specification. A considerable amount of information, especially in regard to the scope and extent of work, is derived from these drawings; also the type of materials and the standards of workmanship to be incorporated into the specification are ascertained from them.

Site investigation and surveys

Some of the information which necessarily has to be included such as soil conditions, site clearance, restrictions on site, water levels, etc. is derived from surveys and site investigation reports. It is normal for a clause to be inserted that the tenderer may inspect, at the engineer's office, a report on a site investigation that has been carried out, although usually such information does not form a part of the contract. The fullest possible information available should be placed at the disposal of the contractor to help reduce the uncertainties for which he must necessarily allow in his tender rates.

Employer's requirements

The requirements of any particular nature of the employer are included in the General Section of the specification. It is often found that access has to be provided to existing works of the employer or to his other contractors for the installation of plant and equipment. In a project the employer may need certain sections of the works to be completed ahead of the others for operational requirements. Likewise, the production process may dictate certain special safety precautions to be taken by the contractor. It is essential that these matters be brought to the notice of the contractor at the time of tender to enable him to assess the responsibilities that he has to bear in relation to these requirements in a realistic manner before the tender is submitted.

British Standards or other international and national standards

It is usual to incorporate British Standards or other international or national standards relating to materials and components into the specification. By so doing the writing in the specification clauses can be maintained as brief as possible, at the same time ensuring a good standard product which is also readily understood by the contractor. Useful documents for the specification writer in this respect are the *Summaries of British Standards* compiled by the British Standards Institution for Building Components, etc., which set out briefly the essential data extracted from British Standards normally used in the building and civil engineering industry.

Reference can also be made to British Standard Codes of Practice in drawing up the specification which are being progressively transformed into British Standards. These ensure that a good standard of construction and workmanship is afforded to the works without the need for lengthy clauses to be incorporated into the specification. Often, where appropriate, international standards or those of other countries are specified.

Standard specifications

Other important sources for the compilation of a specification are standard specifications drawn up by some organizations to suit their particular type of projects. Examples of these are the Department of Transport's Specification for Highway Works (sixth edition – 1986) published by Her Majesty's Stationary Office, the Civil Engineering Specification for the Water Industry (second edition – 1984), published by the Water Authorities Association, and the Specification for Piling together with its companion volume covering contract documentation and measurement (1988), published by the Institution of Civil Engineers in London. Reference to these published documents can always be made in a tailor-made specification in a similar manner in which references are made to British and international standards.

Trade catalogues

Where articles of proprietary manufacture are envisaged to be used on the works the name of the particular manufacturer is specified. It is often necessary to quote the manufacturer's catalogue reference number when an article is produced to many different patterns. It is normal when specifying proprietary products to add the words 'or similar approved' to the particular product specified. By so doing the contractor is in a position to propose a similar article to that named in the specification and incorporate such article in the works, provided the engineer approves the change. By allowing such a flexibility an element of competitiveness is maintained resulting in fair prices for such articles being quoted by tenderers. Again, by this procedure the engineer is absolved from any charge of advertising.

If the fixing or the incorporation of a proprietary article is to be carried out in a special manner required by the particular manufacturer it is usual to specify that the article or product as the case may be should be incorporated in accordance with the manufacturer's recommendation.

Arrangement of specification

The arrangement of the specification and indeed its content should be geared to the type of work envisaged and the type of contract envisaged for the execution of the works. For instance, the philosophy underlying the specification compiled for an all-in or turnkey type of contract will be considerably different from that adopted for an admeasurement type of contract. In the case of the former the design parameters and performance criteria required for the realization of the project will have to be clearly set out in addition to some requirements relating to materials and workmanship and other sections normally provided for an admeasurement contract. In contracts where the design responsibility is transferred to the contractor the degree to which materials and workmanship are specified depends on the extent to which the options available for design may have a bearing on the materials and workmanship that in the event would be required. As an example, if the object of the contract is for a warehouse of a stated capacity to be built, then a contractor could opt to provide the framework for the building to fulfil this object in structural steel, precast reinforced concrete, prestressed concrete or in normal reinforced concrete. Clearly, in such a case specifying the materials and workmanship for the building in a detailed manner can be said to be redundant.

On the other hand, if it has been decided at the outset that this building is to be provided with a structural steel framework, then a specification may be provided for the materials and workmanship that should be adopted for such a building.

When tenders are invited on a design and construct basis it is normal for calculations, drawings, method statements and, if appropriate, detailed specifications in respect of materials and workmanship to be called for from the contractor. The degree to which these requirements are to be prepared and the times at which they should be submitted should be stated in the specification. It is usual to call for such information as is required to define the offer to be provided with the tender, and the balance submitted at appropriate intervals during the currency of the contract, to enable the engineer to carry out such checks as are necessary and supervise the construction of the works adequately.

The majority of civil engineering contracts that are let in the United Kingdom fall into the admeasurement type with the responsibility for the design taken on by the engineer. It follows that for such contracts a design section is not included in the specification. The section headings that are normally adopted in a specification drawn up for these contracts may be stated as follows:

1. General Requirements.
2. Standards and Samples.
3. Materials and Workmanship.

General Requirements

This section of the specification deals with general matters relating to the works. The scope of the works, the location of the site, the availability or otherwise of the services required for the execution of the works, facilities to be provided for the engineer's representative and other contractors together with any special responsibilities that the contractor has to bear in the execution of the works are set out in this section. A typical set of clause headings included in this section may be as follows:

Description of works
Particulars of existing works
Location of site
Access to site and contractor's storage areas
Drawings including list of contract drawings
Setting out
Quality assurance

Soil investigations
Site water and electricity supplies
Assistance to the engineer
Temporary huts
Engineer's site office
Telephones
Photographs
Programme
Method statements and sequence of operations
Weekly and monthly returns
Site records
Temporary works
Temporary buildings, latrines, mess and welfare facilities
Cleanliness of site
Watching and lighting
Advertisements
Attendance upon trades
Safety of works
Legal provisions
General matters affecting the cost of the project
General working requirements.

The requirements stipulated in some of the clauses above, such as provision of engineer's site offices, of telephone and of temporary works particularly required by the engineer, may be paid for as separate items. If this is intended separate items should be inserted as general items in the bill of quantities to cover such requirements to enable prices to be obtained at the time of tender. Alternatively, they may be covered by provisional sum items.

Standards and Samples

The section dealing with standards and samples normally follows the section on general requirements. This section usually contains only a short list of clauses directed at the general aspects of the materials and workmanship specification. A typical list of topics under which clauses are included in this section may be as follows.

General
General clause stating the standards with which the materials should comply (e.g. they should be in accordance with the specification, drawings, bill of quantities and comply with the relevant British Standards unless such standard is amended by the specification drawings or bill of quantities).

Tolerances
A clause stating that the works should be constructed to lines, levels, dimensions on the drawings unless they are ordered to be varied by the engineer.

Samples
A clause setting out the requirements for samples of materials to be employed on the works to be submitted to the engineer for his approval if so directed. The contractor is required under this clause not to use materials that do not correspond with such samples on the works. This clause is also extended often to include workmanship.

Ordering
A clause requiring the contractor to submit names and addresses of makers or suppliers of materials for the engineer's approval before such materials are ordered.

Materials and Workmanship

The third part of the specification is the section covering materials and workmanship, which is crucial for standards of quality of the finished work.

The sequence of setting out the materials and workmanship clauses generally follows one of two forms. One form adopted is for the work to be divided into sections such as earthworks, dredging, concreting, drainage and roadworks and for the materials clauses for each of these sections to be set out first, followed by workmanship clauses. The other form is for all materials required to be specified to be set out in a section of its own followed by another section comprising the workmanship clauses split into sub-sections of work. The form adopted for setting out the materials and workmanship clauses depends on the organization that produces the specification rather than the job for which it is produced. The form chosen is irrelevant: what matters is that the clauses should describe fully the materials and workmanship that are required to achieve the desired quality of the executed work.

The materials and workmanship specification should cover every aspect of the quality of the materials and the standard of workmanship required for the permanent work to be constructed on the works. For a civil engineering contract some of the categories of work included under this section may be listed as follows:

Excavation and earthworks
Underground pipework, drains and sewers
Concrete and reinforced concrete
Steelwork
Roadwork
Miscellaneous items.

Excavation and earthworks arise in almost every civil engineering contract. Although this category of work sounds commonplace, the degree to which this needs to be specified is detailed. A typical specification for excavation and earthwork will cover the following:

Definitions of materials such as
    top soil, suitable material and unsuitable material
Material for backfill
Use of top soil
Removal of suitable and unsuitable materials from site
Disposal of surplus material from site
Re-use of material
Turf and laying turf
Excavation for foundations pits and trenches
Backfill to pits and trenches
Fill
Compaction of fill
Preparation and surface treatment of formations
Keeping earthworks free of water
Dealing with watercourses
Dealing with bad ground conditions and soft spots in
    excavations
Use of explosives.

The above matters relate to excavations and earthworks for general civil engineering contracts. However, if tunnelling, shaft sinking, thrust boring, diaphragm walling or other specialist work is involved the specification will have to cover the requirements for such works as well.

Concrete and reinforced concrete is another category which occurs universally on civil engineering contracts. A typical specification for this category will cover the following:

Portland cement
Other cements

Delivery and storage
Fine aggregates
Coarse aggregates
All-in aggregates
Storage of aggregates
Ready-mixed concrete
Concrete mixes and quality
Preliminary cube tests
Works cube tests
Water
Workability
Machine mixing of concrete
Hand mixing of concrete
Concrete admixtures
Concreting records
Concreting in cold weather
Placing and consolidation of concrete
Curing concrete
Construction joints in concrete
Waterbars
Waterproof membranes
Concrete testing at works
Reinforcement
Concrete finish
Formwork
Removal of formwork
Holes for fitting bolts, etc.
Mortaring and concreting to machinery beds and stanchion
     base plates
Degree of watertightness
Sprayed concrete
Precast concrete.

Again, in specialized work it may be necessary to cover aspects such as grouting, ground treatment, concreting under water, concrete for slipform work, concreting in hot weather and others such as are necessary for the type of use to which the concrete is put.

The detail to which the other categories of work should be specified again depends on the work for which the specification is intended. The specification should be written to achieve the quality of work that is desired: not more, not less.

Chapter 8

Bill of quantities and methods of measurement

A bill of quantities in the context of an admeasurement contract is the document where a brief description of items of work required under the contract together with the quantities of such items of work are listed. The document becomes priced when the tenderers invited to do so insert rates or prices against these items of work, compute the product of the rate and quantity of each item and total them.

The priced bill of quantities forms the basis on which tenders are submitted. Once a contract is formed this document becomes the document on which the contract value is determined and on which financial control is effected.

The bill of quantities forms an integral part of the contract documents together with the other elements of the contract such as the conditions of contract, specification and the drawings.

The bill of quantities may include all or some of the following elements depending on the method chosen for its preparation:

1. List of principal quantities
2. Preamble
3. Schedule of basic rates and prices
4. Sections setting out the work items categorized into:
   4.1 Preliminaries
   4.2 Work items (separately grouped in parts)
   4.3 General items
5. Dayworks
6. General summary.

The fifth edition of the Institution of Civil Engineers' Conditions of Contract (June 1973) (revised January 1979) states at clause 57 that the method of measurement adopted for the preparation of the bill of quantities shall be deemed to be in accordance with the procedure set forth in the Civil Engineering Standard Method of Measurement (CESMM) approved by the Institution of Civil Engineers and the Federation of Civil Engineering Contractors in association with the Association of

Consulting Engineers in 1976, or such later or amended edition thereof. It follows that the current document in use is the second edition of the CESMM published in 1985. Clause 57, however, does not debar the use of a method of measurement other than the CESMM if a statement or the description of the work in the bill of quantities expressly shows to the contrary.

Both the FIDIC (Civil) and the ICE Overseas Conditions of Contract state at clause 57 of the respective General Conditions that 'The Works shall be measured net notwithstanding any general or local custom, except where otherwise specifically described or prescribed in the Contract.' Under these Conditions there is no requirement to adopt the Civil Engineering Standard Method of Measurement as required under the fifth edition of the ICE Conditions of Contract. However, in order to adopt a uniform practice in regard to the method of measurement used in the computation of quantities set out in a bill, and for the same method to be used in measurement of the works, it is usual to adopt a standardized method of measurement, and to state such method in the preamble to the bill of quantities used for such contracts.

Methods of measurement

In March 1933 the Institution of Civil Engineers published a report on engineering quantities which set out uniform principles for drafting bills of quantities for civil engineering works. In 1953 a revised document based on the principles contained in the 1933 report, entitled *Standard Method of Measurement of Civil Engineering Quantities* was published. This was further revised and published in 1963 and a reprint with a metric addendum was issued in 1974.

In 1964 the council of the Institution of Civil Engineers set up a committee to propose revisions to the *Standard Method of Measurement of Civil Engineering Quantities*. In 1971 the work of the revision was placed under the aegis of a steering committee. The revision that resulted was the *Civil Engineering Standard Method of Measurement* published by the Institution of Civil Engineers in 1976 which superseded the earlier document that was in use for a period of some 25 years in the industry.

The principal changes made in the Civil Engineering Standard Method of Measurement (CESMM) were as follows.

1. The greater standardization of the layout and description of bill items prepared in accordance with the previous standard method of measurement;

2. Provision of a systematic structure of bill items leading to uniform itemization and description enabling the use of computers;
3. The sub-division of work items so that they relate more closely to construction costs by the introduction of separate items for method-related charges and other items which are not proportional to the quantities relating to permanent works;
4. The format had been designed to take into account new techniques in civil engineering construction and management, their influence on work itself and the administration of contracts.

Due to the radical new approach adopted, the CESMM did not gain immediate acceptance in all quarters. However, within a relatively short period it was established as the standard method of measurement for contracts governed by the ICE Conditions of Contract. Many consulting engineering, quantity surveying and other organizations involved in the preparation and administration of contracts now tend to have in-house computer facilities for the preparation of bills of quantities and subsequent monitoring of project costs based on the CESMM.

The first edition of the CESMM was in use for almost 10 years before the second edition was published in 1985. The principles of the first edition of the CESMM have been retained in its second edition, which is termed CESMM 2. The measurement notes have been replaced by rules, many of which have the same effect as a former note. These have been categorized and classified for easier reference and interpretation. There is a new section covering sewer-renovation work and sections on ground investigations and rail track have been revised considerably. Many of the changes made are detailed refinements in the light of practice and technological advances that have taken place since the publication of the first edition of the CESMM.

Some promoting authorities have their own method of measurement which includes clauses sensibly similar to those stated in the CESMM or the earlier ICE Standard Method of Measurement with amendments made to suit the local conditions and the particular requirements of the project concerned. Indeed, such documents in the event have been successful where sufficient care had been taken to remove ambiguities in the interpretation of the bill of quantities.

A brief outline of the current *Civil Engineering Standard Method of Measurement* published by the Institution of Civil Engineers is set out below.

Civil Engineering Standard Method of Measurement (CESMM)

The CESMM is the document for measurement which is currently designated at clause 57 of the ICE Conditions of Contract as the document on which the bill of quantities is deemed to have been prepared and measurements made except where any statement to the contrary is expressly stated in the bill of quantities.

The CESMM is intended to be used in conjunction with the fifth edition of the ICE Conditions of Contract. Accordingly, the CESMM refers to individual clauses of those conditions. If the CESMM is to be used with other conditions of contract the provisions of the CESMM have to be carefully checked for compatibility and appropriate amending clauses included in the preamble, which should, also, of course, state that the method of measurement adopted is the CESMM subject to the stated amendments.

There is no basic change in the method of measuring permanent works in the CESMM from the earlier standard method of measurement of civil engineering quantities that has been in use for many years, but the handling of items defined in the CESMM as method-related charges and adjustment items is a major departure from the earlier document. It may be said that the CESMM in fact licenses the inclusion of prices for these two elements in the bill of quantities. It was not unusual for a tenderer when complying with the earlier standard method of measurement to include for method-related charges as sums in the preliminaries bill and to include an adjustment item in the form of tender or as a qualification in the tender letter. It should again be pointed out that if an employer so wishes the CESMM can be still adopted without the facility being afforded to the tenderer for the insertion of method-related charges or an adjustment item, by making a statement in the preamble to the bill of quantities, that the relevant sections in regard to these items in the CESMM do not apply.

The CESMM comprises the following sections:

1. Definitions
2. General principles
3. Application of work classification
4. Coding and numbering of items
5. Preparation of the bill of quantities
6. Completion and pricing of the bill of quantities by a tenderer
7. Method-related charges
8. Work classification.

The definitions given in Section 1 is intended to simplify its text whereby the defined words and expressions can be used as abbreviations for the full definitions: the defined words are also to apply to the terms used in the bill of quantities. At paragraph 1.3 the words and expressions used in the CESMM is defined to have the same meaning as in the ICE Conditions of Contract. The word 'work' is defined to include work to be carried out; goods, materials and services to be supplied, and the liabilities, obligations and risks to be undertaken by the contractor under the contract. This section contains 14 paragraphs devoted to definitions.

Section 2, which deals with general principles, contains seven paragraphs. This section represents a group of rules which set the scene for further and more detailed rules to follow. The CESMM is intended to be used in conjunction with the ICE Conditions of Contract and only in connection with works of civil engineering construction.

The application of the work classification to the bill of quantities drawn up in accordance with the CESMM is outlined in Section 3. The section, among other things, deals with the units of measurement, coding and the manner in which code numbers are to be assigned as item numbers in the bill of quantities. It should be noted that one of the arguments put forward for the adoption of the CESMM is that the work classification and the code numbers which go with it are intended to be used as the core of wider-ranging classification systems. This opens the door for the expansion of the classification system for systematic estimating, cost recording, valuations, financial forecasts and the like with the aid of computers.

The manner in which the coding of items in the work classification can be used for item numbering in the bill of quantities is explained at Section 4. This section comprises seven numbered paragraphs.

Section 5 of the CESMM gives the details of the general rules for preparing the bill of quantities and also deals with the treatment of the elements which are included in the bill such as the daywork schedule, prime cost items, provisional sums and units of measurement.

Paragraph 5.1 points out that the appropriate provisions of this section shall also apply to the measurement of completed work. Paragraph 5.4 deals with the preamble. It is the place for the statement of methods of measurement other than the CESMM adopted on a particular contract, any modifications to the CESMM or the rules of measurement for any contractor-designed work. The preamble should also include the requirements derived from paragraphs 5.5, 5.20, 6.4 and 7.7 of the CESMM. Paragraph 5.5 states that where excavation, boring and driving is included in the work, a definition of rock shall be given in the preamble which should be used for purposes of measurement. Paragraph 5.20 makes it a requirement for the presence of bodies of open water (other than groundwater) on the site or at the boundary of the site to be mentioned in the preamble. The CESMM does not advocate the inclusion of clauses in the preamble which are already covered in the Conditions of Contract and in the appendix to the form of tender, such as, say, 'the cost of items against which no rate is entered is deemed to be included in other prices or rates entered in the bill of quantities'; this requirement is covered under clauses 11(2) and 55(2) of the Conditions of Contract. Section 5 of the CESMM contains 27 numbered paragraphs.

The manner in which the bill of quantities shall be completed and priced by the tenderer is stated in five paragraphs at Section 6. The manner in which the adjustment item given in the grand summary is to be assessed for interim certificates is set out in this section. It should be noted that the adjustment item is deemed a lump sum which in the event is payable in whole irrespective of the contract value being greater or less than the tender total. It is a requirement under paragraph 6.4 that the method of payment applicable to the adjustment item appears in the preamble.

The rules governing the use of method-related charges are given in eight numbered paragraphs at Section 7. The description of items and prices in respect of method-related charges are to be inserted at the tenderer's option.

The only direction given in this regard is that, where possible, the itemization of method-related charges should follow the order of classification and the other requirements set out in class A of the work classification, distinguishing between time-related charges and fixed charges. There is a provision, however, to the effect that method-related charges may be inserted to cover items of work other than those set out in class A of the work classification. There are some features relating to method-related charges set out in the rules which are worth special mention. The insertion by the contractor of an item for a method-related charge in the tender does not bind him to adopt that method during the execution of the works. These charges are not subject to admeasurement. Under paragraph 7.7 payment for such charges is to be made, pursuant to clauses 60 (1) and 60(2)(a) of the ICE Conditions of Contract and a statement to this effect is to appear in the preamble. If the method is not adopted in whole or in part the contractor is still entitled to payment of the method-related charges, and payment in the event is to be on the basis set out in the CESMM for the adjustment item.

The work classification is set out in Section 8. This is the section that will be used most frequently for the preparation of the bill of quantities.

The classification divides the work covered by the CESMM into 25 classes lettered A to Y. Each class contains a statement of inclusions and exclusions, a classification table and rules.

The inclusions and exclusions statement gives the general types of works which are included in a class and those which are excluded from that class. These lists are important to the interpretation of the coverage of the bill items generated by the classes, but do not set out to be comprehensive.

The classification table tabulates the work components covered by a class divided into three divisions.

The measurement rules set out alongside the classification table on the right-hand pages are as important as the classification tables themselves. These define, among other things, the parameters for measurement of the items in the bill of quantities covered by the respective class of the work classification.

The classes of work covered by the work classification are:

Class A: General items
Class B: Site investigation
Class C: Geotechnical and other specialist processes
Class C: Demolition and site clearance
Class E: Earthworks
Class F: *In situ* concrete
Class G: Concrete ancillaries
Class H: Precast concrete
Class I: Pipework – pipes
Class J: Pipework – fittings and valves
Class K: Pipework – manholes and pipework ancillaries
Class L: Pipework – supports and protection to laying and excavation
Class M: Structural metalwork
Class N: Miscellaneous metalwork
Class O: Timber
Class P: Piles
Class Q: Piling ancillaries
Class R: Roads and pavings
Class S: Rail track
Class T: Tunnels

Class U: Brickwork, blockwork and masonry
Class V: Painting
Class W: Waterproofing
Class X: Miscellaneous work
Class Y: Sewer renovation and ancillary works.

Bill of quantities

The elements included in the bill of quantities are described in the following terms.

List of principal quantities

The inclusion of a list of principal quantities in a bill is a requirement dictated by the CESMM. At paragraph 5.3 the purpose for the list is defined. The requirement is to set out a list of principal components of the works with their approximate quantities. This is given solely for the benefit of the tenderer to make a rapid assessment of the general scale and character of the proposed works prior to the examination in depth of the documents comprising the tender. The CESMM does not give guidance on which quantities should be classed as principal ones. In a contract covering an industrial undertaking it can be said that about 70 per cent of the total cost may be covered by costs relating to excavation, concrete, formwork, reinforcement, prime cost items and provisional sums. Therefore for such an undertaking the principal quantities given may be confined to setting out the quantities relating to these elements in each part of the bill. If the project is confined to one site where the conditions are generally similar in respect of the work included in each part, the subdivision of the principal quantities into parts may not be required. Depending on the type of project, the elements that bear a major portion of the total cost should be identified and set out together with the estimated quantities relating to such elements.

Preamble

The preamble to the bill of quantities embodies the instructions and references to the obligations intended to be imposed by the document on which the tender is to be based. The main purpose of the preamble is to assist the tenderer to price the items in a bill without ambiguity. The preamble also serves as an instrument which considerably assists the engineer in applying uniform criteria for measurement of quantities of work done during the progress of the contract.

When the CESMM is adopted for the preparation of a bill of quantities and for the measurement of works the function of the preamble in the context of the bill of quantities is limited considerably. The CESMM only defines four functions of the preamble. These are for the statement of any special methods of measurement adopted for the particular contract; the manner in which the adjustment item and method-related charges, if included by the tenderer in the bill, are to be certified and paid, and the fourth function is that the preamble should identify bodies of open water on or bounding the site, such as a river, stream, canal, lake or a body of tidal water. The CESMM, being designed for use with the ICE Conditions of Contract, relies on the definitions and interpretations therein together with the definitions and rules in the the CESMM to cover clauses in the preamble which were hitherto normally provided under the earlier ICE Standard Method of Measurement.

Where the FIDIC (Civil) Conditions of Contract, the ICE Overseas Conditions and standard forms of contract other than the fifth edition of the ICE Conditions of Contract (June 1973; revised January 1979) and the ICE Conditions of Contract for Ground Investigation are adopted, the method of measurement used for the preparation of the bill of quantities must be stated in the preamble. This may be effected by naming in the preamble a standard method of measurement such as the CESMM or other standard method with modifications and amplifications as are necessary thereto, or simply by listing in the preamble the rules under which the bill of quantities was prepared and the work is to be measured. In order to illustrate the elements that should be incorporated into such a preamble it may not be inappropriate to list certain directions which were considered necessary to be given to tenderers in the ICE standard method of measurement which was in use before the publication of the CESMM. These are set out below:

1. Attention is directed to the form of contract; the conditions of contract, the specification and the drawings and these documents are to be read in conjunction with the bill of quantities.
2. The bill of quantities has been drawn up in accordance with the

Standard Method of Measurement of Civil Engineering Quantities published by the ICE.

3. The prices and rates to be inserted in the bill of quantities are to be the full inclusive value of the work described under the several items, including all costs and expenses which may be required in and for the construction of the work described, together with all general risks, liabilities and obligations set forth or implied in the documents on which the tender is based. Where special risks, liabilities and obligations cannot be dealt with as above, then the price thereof is to be separately stated in the item or items provided for the purpose.

4. A price or rate is to be entered against each item in the bill of quantities whether quantities are stated or not. Items against which no price is entered are to be considered as covered by the other prices or rates in the bill.

5. Any special methods of measurement used are stated at the head of or in the text of the bill of quantities for the trades or items affected. All other items are measured net in accordance with the drawings and no allowance is made for waste.

6. General directions and descriptions of work and material given in the specification are not necessarily repeated in the bill of quantities. Reference is made to the specification for this information.

In practice the preamble is so drafted as to incorporate these directions, amplified as appropriate to enable the contractor to build up his rates for the various items included in the bill.

In amplification of these directions some of the matters that are usually covered in the preamble are:

1. Programming and execution of work by nominated sub-contract, provisional sums and P.C. items.

2. The content of work to be included in the percentage quoted for adjustment of provisional sums for work envisaged to be carried out by nominated sub-contractors or suppliers.

3. Definition of words used to describe items in the bill, such as 'supply', 'handle', etc.

4. Clarification of the basis of payment for various items in the bill, such as for excavation, fill, concrete, formwork, reinforcement, brickwork, pipes, etc.

5. Reference to the manner in which works carried out in accordance with the daywork bills will be paid. The clauses are drawn up to suit the method adopted for valuation of daywork chosen from the provisions laid down in clause 14 of the Standard Method of Measurement.

6. The manner in which rates will be fixed for additional and extra works.

Schedule of basic rates

The schedule of basic rates and prices, which is also termed a schedule of basic prices, is sometimes included between the preamble and the bill in a contract document prepared for works of a civil engineering nature.

The schedule of basic rates and prices contains spaces for the tenderer to fill in the basic rates for the labour, materials and plant used in the preparation of his tender. The schedule requires the make-up figures of the ingredients that form the bulk of the rates quoted by the tenderer to be stated with the tender. Basic rates are accordingly asked for in respect of labour, plant and materials such as aggregates, cement, reinforcement, timber, pipes, bricks, etc. as appropriate to the particular works for which the contract is framed.

The schedule has two possible functions in the context of an admeasurement contract. The first is as a basis for the fixing of new schedule rates effected in the process of rate fixing; second, it may be used for the contract price variation covering increases or decreases as the case may be in statutory or unavoidable changes in prices during the period of the contract; this only applies where a fixed price contract does not obtain.

Pricing sections

The pricing sections of the bill of quantities comprise lists of items identifying descriptions and estimated quantities of work envisaged in the contract. Spaces are provided for the tenderer to enter his rate for carrying out each of the items described. The rates and quantities are extended by the tenderer which result in the total price for each of the items so described. The rates quoted are in effect inviolate, and are only subject to genuine errors which are recognized as such by the engineer before the award of a contract. The quantities, on the other hand, are estimated and the price against each item varies, depending on the actual quantities measured during the currency of the contract.

Provision exists for some of the items to be quoted as a lump sum together with provisional sums to be included for items which are not defined at the time of tender.

The sections put out to a tenderer for pricing a tender in respect of works envisaged to be paid on an admeasurement basis are set out below.

Preliminaries
If the CESMM is adopted a section (or bill) for the inclusion of preliminaries is not required. However, provision exists in a bill of quantities prepared in accordance with the CESMM for items traditionally included as preliminaries to be listed and priced in the General Items section of the bill of quantities.

In the preliminaries section of a bill of quantities certain items arising from the general obligations of the contractor under the conditions of contract and the specification are listed and the opportunity is given to the contractor to price them. In some bills the clause numbers are inserted by the engineer beforehand, while in others, blank spaces are provided in the bill to enable the contractor to enter the clauses for which he wishes to price in respect of preliminaries.

Clauses 8 and 9 of the earlier ICE Standard Method of Measurement state that the general obligations as defined in the conditions of contract and the cost of temporary works as a general principle should be covered by the billed rates. Therefore strictly there is no need for a separate preliminaries bill to be provided in civil engineering contracts.

In practice, however, on large contracts, especially those awarded for construction work overseas, the inclusion of a preliminaries bill is more the rule than an exception. This bill gives the contractor an opportunity for pricing for items such as constructional plant, advances and for certain mobilization costs which have to be incurred.

Work items (separately grouped into parts)
The work items in the bill of quantities form the bulk of the bill. The arrangement of the work items into numbered bills or parts (if based on CESMM) should be carried out according to the location of individual parts of the works in the general scheme or according to the character of the works performed. The manner in which items are grouped into a bill depends largely on the type, character and extent of the works.

In grouping the bills on an individual project the logical pattern that would be used would be by location of individual structures. If extensive site work is envisaged this would be the first item to be undertaken and consequently form the first bill. In a site investigation contract the first bill may be for the mobilization of plant and equipment followed by site work in trial borings, and thereafter the bill for laboratory testing of soil samples will be included. If the methods of construction envisaged for different parts of the work are likely to be significantly different then it would be rational for work in such parts to be grouped separately. At all events, what is required is for a logical sequence to be followed in grouping the work items so as to assist the tenderer during pricing and to assist all persons using the bill in the financial control of the project. A well-grouped bill of quantities makes planning and financial control straightforward, and facilitates prompt measurement of quantities and the settlement of the final account. Each item of work listed in a part of the bill is either consecutively numbered or if the CESMM is adopted the items are coded.

General items
The general items that are included in a bill of quantities are those which are classified as Class A in the CESMM work classification or those which flow from clause 9 of the earlier ICE Standard Method of Measurement if the latter document or a document based on it is adopted for the preparation of the bill of quantities.

The CESMM makes use of general items to include preliminary items which were earlier traditionally included in the 'preliminaries' bill. The CESMM in the general items part of the bill aims to include in it a group of items which are general in that they are not directly related to the permanent works. This section is to include items for general obligations imposed by the contract, site services and facilities, temporary works, testing of materials and work, provisional sums and prime cost items. These items are to cover elements of the cost of the work which are not considered as proportional to the quantities of the permanent works. It follows that the general on-costs are not spread over the rates or prices included in this section. The sub-headings of the groups of items that are required to be included under general items where the CESMM is adopted are:

Contractual requirements
Specified requirements
Method-related charges
Provisional sums
Nominated sub-contracts which include work on the site
Nominated sub-contracts which do not include work on the site.

It should be noted that the items for method-related charges stated above, if any, are to be inserted by the tenderer distinguishing between time-related and fixed charges; a space is left in the bill for the tenderer to insert such items. All other items are inserted by the engineer in this part of the bill with only spaces left for pricing such items, in a manner similar to that adopted in other parts of the bill of quantities.

Dayworks

The ICE Conditions of Contract at clause 52(3) empower the engineer to order additional work or substituted work to be executed on a daywork basis if, in his opinion, it is considered necessary or desirable to do so. Similar provision exists for ordering work on a daywork basis in both the FIDIC and Overseas Conditions of Contract. For a contract designed for the execution of civil engineering works provision is made inevitably for some miscellaneous work to be carried out on a daywork basis. Work ordered to be carried out on a daywork basis is incidental to contract work and accordingly the rates and prices included in a contract for such work assume this position.

The ICE Conditions of Contract stipulate that the contractor shall be paid for work ordered to be carried out on a daywork basis under a daywork schedule included in a bill of quantities at the rates and prices affixed to it by the tenderer and, failing the provision of a daywork schedule, that payment be made under rates and prices and conditions contained in the 'Schedule of Dayworks' (carried out incidental to contract work) issued by the Federation of Civil Engineering Contractors current at the date of the execution of daywork.

In drafting a bill of quantities for works to be executed in the United Kingdom it is quite usual to adopt the schedule of dayworks carried out incidental to contract work, issued by the FCEC especially in regard to plant used on daywork, as the rates for a comprehensive list of plant are included in this schedule. The rates for labour are likewise required to be based on this schedule which, in turn, refers to the rates of labour listed in the Working Rule Agreement of the Civil Engineering Conciliation Board for Great Britain, or that of other appropriate wage-fixing authorities, where such recognized wage-fixing authorities exist.

In the FCEC schedule of daywork the percentage additions to be made to the basic costs are stipulated in so far as labour and materials are concerned. The elements of cost that are deemed to be included in the percentage addition are also defined. In practice it is quite usual for the percentage addition to be left open for the tenderer to fill in. In doing so, the element of costs that are to be included in the percentage addition is sometimes defined in the preamble in a manner different from the definition made in the FCEC document. This is usually done to ease site administration. For instance, in respect of labour the percentage addition to be inserted by the tenderer often is to include for all elements of cost other than the basic rate of pay and the plus rates for skill only; in the event, the payments that have to be made to workmen according to the Working Rule Agreement, such as the joint board supplement, guaranteed bonus, plus rates for conditions and the like, together with the costs in connection with incentives the contractor may have to extend to his workmen, are all required to be included in the percentage addition.

It is therefore usual for the percentage addition quoted by the tenderer to be in excess of the percentage stipulated in the FCEC schedule of daywork.

For contracts carried out overseas it often becomes necessary to include a daywork schedule listing in it the main categories of plant and labour in the bill of quantities. This requirement arises when appropriate standard schedules are not available in the countries where the works are required to be constructed. The schedules attached to the tender in such cases are priced by the tenderer for work to be executed on a daywork basis.

The manner in which financial provision is made for daywork in a bill of quantities is by inclusion of three separate provisional sums for labour, materials and plant. Following each of these provisional sums, items are inserted for the tenderer to quote the respective percentage adjustment required by him for execution of these works.

The CESMM deals with its rules on daywork schedules at paragraph 5.6 and sets out the manner in which provisional sums for daywork should be incorporated into the bill of quantities at paragraph 5.7. Daywork included as provisional sums falls under Class A of its work classification and is included as 'General Items'. The earlier ICE Standard Method of Measurement provides at clause 14 for the valuation of work carried out on a daywork basis to be made as follows:

1. By a daywork schedule prepared in such a way as to enable entry in detail of separate rates for the respective classes of labour, materials supplied and the hire of plant; such rates to cover overhead charges and profit, site supervision and staff, insurances, holidays with pay, use and maintenance of small hand tools and appliances (but not sharpening of tools), non-mechanical plant and equipment such as ladders, trestles, stages, bankers, scaffolding, temporary track, wagons, skips and all similar items, unless these are used or set up exclusively for daywork, and in the case of rates for mechanically operated plant coming under the heading of 'plant', consumable stores,

fuel and maintenance. When travelling allowances or travelling costs (transport of men by contractors' transport), lodging allowances and any other emoluments and allowances payable to the workmen at the date of submission of the tender, are included it should be so stated in the preamble, or

2. By using (in whole or part) the current schedules of dayworks carried out incidental to contract work issued by the Federation of Civil Engineering Contractors, or

3. By adapting the current schedules of dayworks carried out incidental to contract work issued by the FCEC in such a way as to enable entry of percentages for labour and materials differing from those given in the schedules of dayworks, and in the case of plant, percentages varying the hire rates given in the schedule of dayworks.

General summary

The general summary or the grand summary of the bill of quantities is the place where the respective parts or bills are tabulated with provision for insertion of the total amounts brought forward from such parts or bills. The CESMM calls for a provisional sum for contingencies, if required, to be included in the general summary together with a space for the insertion of an amount as the adjustment item following the total of the amounts brought forward from the part summaries. In a bill of quantities which is not prepared in accordance with the CESMM the provisional sum for contingencies may be included either in the General Items bill or in the general summary in accordance with the individual preference of the engineer responsible for drafting the contract documents.

The tender total is the total of the amounts tendered for each of the parts tabulated in the grand summary including the items provided, if any, in respect of contingencies and the adjustment item. This tender total is defined in the ICE Conditions of Contract as the total of the priced bill of quantities at the date of acceptance of the contractor's tender for the works. The form of tender of the ICE Conditions of Contract does not contain a space for the insertion of the tender total; the form, however, refers to it. Both the FIDIC and the Overseas forms provide a space in the 'Form of Tender' for the insertion of the tender total in their respective forms of tender.

Provisional sum items

Provisional sum items are those for which sums of money are provided in the bill of quantities for contingencies, additional or extra works, or for cost of works envisaged to be carried out on the basis of nominated sub-contracts. These sums are to be expended only if an occasion arises and only at the direction and discretion of the engineer. The work executed is measured and valued at the rates (and/or analogous rates) contained in the priced bill of quantities or as prime cost items or on the basis of dayworks all in accordance with the provisions relating to such works in the conditions of contract.

Provisional items

Provisional items are items marked (Provisional) in the text of items in a bill of quantities provided in the bill for work which may or may not be undertaken. Examples of provisional items could be 'extra over items for excavation' where in rock, 'making good soft spots under foundations' or, say, for the 'import of hoggin fill for road foundations' where the excavated material from the site may in the event be found unsuitable. These eventualities may not manifest themselves until the actual work is undertaken, and all that is possible to do during the preparation of the bill of quantities is to make a reasoned assessment of the possible quantity and term the items provisional. It must be pointed out that neither the fifth edition of the ICE Conditions of Contract nor the CESMM define provisional items. It follows that when such items are inserted that they should be separately defined in the preamble to the bill of quantities. Such items are, in effect, provisional sum items which may be used in whole or in part or not at all at the direction and discretion of the engineer.

Prime cost items

The term 'prime cost' (P.C.) may be defined as the net sum entered in the bill of quantities by the engineer as the sum provided to cover the cost of, or to be paid by the contractor to merchants or others for specific articles or materials to be supplied or work to be done.

When prime cost items are entered in the bill of quantities provision is to be made in the bill to permit the tenderer to fill in separate sub-items for the following:

1. Charges and profit on prime cost as a percentage thereof; and
2. An item for handling and fixing connected with the item for which the prime cost is intended.

The tenderer is required to compute the amount derived by the application of the percentage quoted by him to the prime cost sum entered by the engineer and insert this sum in the amount column of the bill.

The amount payable in the event shall be the actual cost of the item for which the prime cost is intended, plus the percentage quoted by the tenderer of the actual cost plus the fixed sum in the tender for the handle and fix element of the prime cost item.

Chapter 9

Insurances

Principles of insurance

Insurance is a special type of contract under which a sum of money will be paid by the *insurer* on the happening of a specified event. The person to whom payment is made is the *insured*. It is, of course, uncertain whether the specified event will actually happen, and this uncertainty, in the context of the likelihood of the event happening and the consequences if it does, may be termed the *risk*. In calculating the premium that he requires to enter into the contract the insurer will be very much concerned in assessing that risk. Where the event is certain to take place, although there may be uncertainty as to when it will take place as in a policy securing payment on a person's death, the matter becomes one of *assurance* not *insurance*.

Although it is possible to insure very many risks in today's insurance market, not all risks can be insured. In the United Kingdom insurance is subject to a set of legal rules as well as certain practices which stem from commercial and moral considerations.

Before any insurance can be taken out it is a prerequisite that the insured must have an insurable interest in the subject matter of insurance. Basically this means that there has to be a legally recognizable relationship between the insured and that which is being insured, whereby he benefits from its continued safety or freedom from liability and would be prejudiced by its damage, destruction, or the creation of liability. Insurance cannot be used as a device to cover speculative risks or for gambling.

The insurance contract is a legally binding agreement between two or more parties. There is no legal requirement for insurance contracts to be in writing unless they relate to certain aspects of marine insurance or are contracts of guarantee. However, in practice the insurance policy provides written evidence of the contract. Insurance contracts are different from normal legal contracts in that they are contracts of 'utmost good faith' (*uberrimae fidei*) as against 'good faith' which has to be observed in all legal contracts. The common law principle applicable to most commercial contracts, such as a contract for sale, is that of *caveat emptor* (let the buyer beware). In contracts for sale of an article the seller must not misrepresent the article or deceive the buyer but he need not point out all the defects inherent therein. Consequently, should a defect be found after the sale of such an article the buyer has no right of action at common law (as distinct from statute) against the seller, unless a warranty on the non-existence of such a defect was given or representations were made by the latter.

Contracts of insurance are based upon mutual trust and confidence between the insurer and the insured. The insurer has to rely mainly on the information given by the proposer for the assessment of the risk. The exercise of utmost good faith legally obliges each party to a proposed contract to reveal to the other all information that would influence or affect the other's decision to enter into a contract whether or not such information is requested. This duty is imposed on both the proposer, who is the prospective insured, and on the insurer. However, the duty bears more heavily on the proposer, because he knows or should know all about the risk put forward for insurance. In discharging this duty the proposer has to divulge to the insurer all material facts which may influence the insurer's assessment of the risk; merely answering truthfully the questions asked in the proposal form may not suffice.

The duty to disclose all material facts continues until the contract is entered into. It follows that if there is any change in those facts after the proposal form has been completed but before the contract is concluded those changed facts must be notified to the insurer. Most insurance policies require as a condition of the policy the notification by insured of any changes affecting the risk during the currency of the policy. The proposer is, however, not penalized for not revealing facts which he does not know or which he cannot reasonably be expected to know. Again, because the principle of utmost good faith is applied to prevent the parties from being drawn into a contract in the ignorance of facts which might enhance the risk, the following categories of facts need not be disclosed:

1. Facts that may diminish the risk.
2. Facts that are known by the insurer.
3. Facts capable of discovery by the insurer from information supplied.
4. Matters of law.
5. Facts that are not noticed by an insurer's representative at a survey, provided that the necessary details are not withheld or concealed from such a representative by the proposer.
6. Facts that are superfluous because they are already covered by express or implied warranties.

The information supplied by the insured is the basis of the contract of insurance, and therefore if in doubt all known information should be disclosed.

Types of insurers

The different types of insurer available for placing insurance in the UK market, which transacts a considerable amount of international insurance, may be stated to be as follows:

1. Lloyds underwriters
2. Proprietary companies
3. Mutual companies
4. Friendly societies
5. Mutual indemnity associations
6. Captive insurance companies.

The bulk of insurance placed in the United Kingdom falls into the first three categories of insurer listed above.

The Corporation of Lloyds traces its origins to a coffee house in London owned by Edward Lloyd in the seventeenth century. This coffee house was situated near the Thames, like many others, and these were places where wealthy merchants met. In those days ships and cargoes were insured by a person who became known as a broker approaching merchants for subscriptions to take on a share of risk for a portion of the premium. Lloyd himself did not transact any insurance business but merely provided reliable shipping information and excellent facilities for doing so. Thus his coffee house became a recognized place for obtaining marine insurance in London. To this extent the Corporation of Lloyds even to this day performs the same function; it provides the facilities for insurance to be transacted but does not carry out insurance itself, One does not insure with Lloyds but does so at Lloyds.

Insurance at Lloyds is transacted by individual underwriting members who are grouped together in syndicates and are personally liable for the business they underwrite to the full extent of their means. Lloyds was incorporated by an Act of Parliament in 1871. It is governed by a committee elected from members which lays down very stringent requirements for membership. Insurance at Lloyds is always conducted through Lloyds' brokers who, at the outset, have to satisy stringent requirements as to their integrity and financial standing.

A proprietary insurance company is a company owned by its shareholders. The shareholders have limited liability and are incorporated by Royal Charter, by special Act of Parliament, by registration under the Companies Act or, in the case of some of the older companies, by a deed of settlement. The privilege of gaining incorporation by Royal Charter is only enjoyed by a few companies; also those incorporated by a Special Act of Parliament are fairly rare. Proprietary companies deal direct with the public and compete freely with other categories of insurer.

Mutual companies are those owned by policy-holders who share any profits that are made. Some of these companies specialize in life assurance. These companies are incorporated by registration under the Companies Act or were those created originally by the older form of deed of settlement; most in the latter category have since registered under the Companies Act.

Friendly societies are concerned only with industrial life assurance. These societies have to be registered, and it is a criminal offence to carry on such insurance business in contravention of this provision.

Mutual indemnity associations were originally formed as trade associations as a club for members within a trade. These have since widened their scope by opening their membership to the public. These associations are mainly seen in the marine insurance sphere in connection with liability cover.

Self-insurance is the term used when an organization creates its own fund to meet possible losses. A sophisticated form of self-insurance is sometimes provided by large corporations and multinational organizations through the establishment of a captive insurance company to carry out the insurance business of the respective parent group. Any captive insurance company established in and carrying on the business of insurance in the United Kingdom requires the approval of the Secretary of State, like any other insurance company.

Types of insurance

The main classes of insurance that are effected nowadays may be broadly classified into marine, life, accident, motor, fire, engineering and aviation. More and more types of risks are covered by the insurance industry today, and with the ever-increasing pace of development the nature of risks covered as well as the volume of business transacted is increasing. Almost all classes of insurance can be transacted at Lloyds or with the larger and better-known proprietary companies. Some of the mutual companies specialize in life assurance and the classes of insurance transacted by friendly societies and mutual indemnity associations are limited.

Assessment of risk

It is axiomatic that for insurance to be effected it must be possible to calculate the risk of a loss. Most risks can be calculated mathematically. Past occurrences of losses, and statistics maintained on them, act as the basis for the prediction of the probable occurrence and extent of a loss. If it is not possible for the probable extent of a loss to be reasonably predicted it will be very difficult to obtain insurance. Insurers have over the years compiled statistics and data for the calculation of risks pertaining to many classes of insurance. The basic parameters considered in assessing a risk are the probability of the occurrence of damage and the corresponding magnitude of the loss caused by such an occurrence. To illustrate this in simple terms: if the probability of occurence of damage is 1 in 100 and the expected magnitude of loss incurred is 25 per cent of the amount of the insured sum, then such a risk will be rated as the product of the probability and the percentage of the loss which would be 0.01 times 25 per cent, which equals 0.25 per cent of the insured sum. It follows that if the insured sum is £100 000 the risk rating is £250. This value becomes the net premium; the cost to the insured will be the gross premium which includes the insurer's overhead costs and commissions to brokers or agents. The mark-up to cover the latter costs depends to a large degree on the class of insurance, the type of insurer, and also at times on the credibility of the insured, but it could be up to 50 per cent of the amount of the net premium. Premiums are, however, frequently negotiated with insurers on the basis of their educated judgement rather than as a pure mathematical exercise.

Insurance cover required under civil engineering contracts

Under the majority of standard forms of contract adopted for the construction of civil engineering works, such as the ICE, Overseas and FIDIC, it is a requirement that certain classes of insurance be taken out by a contractor. These insurances are in respect of the corresponding risks and liabilities that have to be basically borne by the contractor. Although such insurance is primarily for the benefit of the contractor, it also serves as a financial security for the employer. The presence of satisfactory insurance is a form of guarantee to the employer that his contractor will not be financially embarrassed or be forced to go into liquidation by the occurrence of an insured risk to the detriment of the fulfilment of the work for which he was engaged. The classes of insurance required to be taken under these conditions of contract fall into two categories, namely, works and liability insurance policies.

Insurance of works

The ICE Conditions of Contract at sub-clause 20(1) require the contractor to take full responsibility for the care of the works from their commencement to 14 days after substantial completion. Further, he is required to exercise this responsibility and care to any outstanding work undertaken by him to finish during the period of maintenance until such outstanding work is complete. Under sub-clause 20(2) the contractor is required to repair and make good at his own cost any damage, loss or injury from any cause whatsoever (unless they are excepted risks) that shall happen to the works while the contractor is responsible for the care of the works.

Under clause 21 of the conditions of contract the contractor is required to insure the permanent and temporary works, materials on-site and constructional plant to their full value against loss or damage for which the contractor is responsible under the terms of the contract. This insurance is to be effected in the joint names of both the employer and the contractor. It should be noted that the contractor is not obliged under this clause to insure against the necessity for repair and reconstruction of any work due to defective materials or workmanship unless the bill of quantities provides a special item for such insurance. The contractor is, however, expressly liable to repair any damage arising from defective materials and workmanship under the terms of sub-clause 39(1)(c) during the progress of the works and under sub-clause 49(3) during the period of maintenance. Policies of insurance, however, cover actual losses or damage suffered as a result of a defect in workmanship or materials, although the cost of replacement or rectification arising from such a defect is not insured. In other words, if a wall which had been incorrectly built falls over and damages the adjoining property, the cost of rebuilding the wall would be excluded, but the damage to the adjoining property would be covered. The object of insurance is to give protection against the fortuitous happening and does not in any way attempt to guarantee that a contractor will necessarily construct his works to specification.

The excepted risks for which the employer is responsible and consequently do not require to be insured under the terms of clause 21 of the ICE Conditions of Contract are:

1. Riot.
2. War, invasion, act of foreign enemies' hostilities (whether war is declared or not), civil war, rebellion, revolution, insurrection or military or usurped power.
3. Ionizing radiation, contamination by radioactivity or from any nuclear fuel or from any nuclear waste from the combustion of nuclear fuel, radioactive toxic explosive or other hazardous properties of any explosive nuclear assembly or nuclear component thereof.
4. Pressure waves caused by aircraft or other aerial devices travelling at sonic or supersonic speeds.
5. A cause due to use or occupation by the employer, his agents, servants or other contractors (not being employed by the contractor) of any part of the permanent works.
6. Fault, defect, error or omission in the design of the works (other than a design provided by the contractor pursuant to his obligations under the contract).

Many of the above risks are commercially uninsurable: those which are insurable tend to attract high premiums.

Riots are insurable apart from in certain areas such as Northern Ireland and other countries where riots occur frequently. War risks are of such a magnitude that they are not insurable (in the context of civil engineering) on a commercial scale; these risks become a risk of the government. Nuclear risks are also uninsurable commercially; these are also risks for which a government is in theory responsible. Damage arising out of pressure waves caused by aircraft or aerial devices is also normally uninsurable; in the United Kingdom the government originally undertook to compensate damage resulting from the Concorde flights. Causes due to occupation of the works by the employer, his agent or other contractors are insurable by the employer. Losses arising from a design defect are normally covered by the designer's professional indemnity cover. Obviously the contractor cannot be asked to take on the responsibility for a design for which he is not responsible. It has to be pointed out that the task of determining whether a fault is due to defective design, workmanship or negligence is not an easy one. A thorough investigation has to be undertaken and considerable care should be exercised in reaching the correct conclusion in this regard.

The insurance requirements flowing from clause 21 of the conditions of contract are met by taking out what is termed a contractor's all risk (CAR) policy. Such a policy protects the property of the insured for which he is responsible and indemnifies him against a loss or damage arising from whatsoever cause to the works, materials and constructional plant. A contractor's all risk policy can be taken out in such a manner that the employer and the contractor can be covered during the performance of the works and during the period of maintenance. The requirements of clause 54(3)(b) of the ICE Conditions of Contract for insuring goods and materials stored off-site where such are to be paid for by the employer can also be covered under a contractor's all risk policy. The contractor's all risk policies that are taken out exclude the excepted risks and also generally do not cover certain categories such as:

1. Costs of remedying the breakdown of plant and machinery by reason of fair wear and tear and gradual deterioration thereof.
2. Loss or damage to motor vehicles, locomotives, marine craft and aircraft. These risks are usually covered by specific insurances such as motor, engineering, marine and aviation. Special arrangements can be made where certain types of insurance can be covered by this policy.
3. Loss or damage due to confiscation, nationalization or compulsory acquisition by the government, public, local or competent authority.
4. The cost of replacement or rectification rendered necessary by defects in materials and workmanship other than liability for consequent damage arising therefrom.
5. Liquidated damages and other costs incurred by the contractor for non-compliance with the contract.
6. Silting up of dredged areas or loss or damage to underwater excavation, rock protection and the like caused by normal tide and current.
7. Excesses applicable in respect of each occurrence giving rise to loss or damage.

Public liability insurance – damage to persons and property

Clause 22(1) of the ICE Conditions of Contract requires the contractor to indemnify the employer against any losses and claims for injury to persons or damage to property arising out of or in consequence of the construction and maintenance of the works. The provisos to the clause may be summarized as follows:

1. The contractor's liability to indemnify the employer as required above is reduced proportionately to the extent that it is contributed to by the employer, his servants or agents.
2. The contractor is not liable for claims arising out of damage to crops on-site, interference with rights of way, loss of such amenities and damage which is the unavoidable result of the construction of the works in accordance with the contract.
3. The contractor is not liable for claims for injuries or damage arising from acts done or committed by the engineer or the employer, his servants, agents or other contractors who are not employed by the contractor.

Under clause 22(2) the employer indemnifies the contractor from all claims arising out of the matters referred to at (1) and (2) above and in respect of (3) above, to the extent the contractor has not contributed to claims arising therefrom.

Clause 23 requires the contractor to insure against the liabilities imposed upon the contractor under clause 22 and the policy is required to include a provision whereby the insurer indemnifies the employer against claims made on him, for which the contractor is entitled to an indemnity under the policy. The insurance is to be effected for at least the amount stated in the appendix to the form of tender; various employers place differing amounts as the limit. It has to be borne in mind that this is an area where the contractor is most likely to be underinsured. An accident occurring within the works can cause damage over and above the value of the works to adjoining properties. It follows that the cost of the works has no direct relationship to the amount for which public liability insurance should be taken. The risk attached to the methods of construction that are adopted and the nature and types of structures that exist on the adjoining properties would have a greater influence on the cost involved. Generally it is considered in the industry that this insurance should be taken for a minimum of £1.5 million for any one incident if such works are constructed in the United Kingdom. Depending on the location, this limit may have to be increased where appropriate.

Although clause 23 imposes a requirement for insurance of the contractor's obligations it does not do so in respect of the indemnity given by the employer in respect of his obligations at clause 22(2). Although the contract does not impose an insurance requirement on the employer in respect of his obligations, if full third-party protection is to exist on the contract such insurance should be taken out by the employer to the extent that it is not excluded by insurers.

Public liability insurance policies, in common with materials damage policies such as the contractor's all risk policy, exclude cover for certain categories of insurance. Such a policy excludes risks that are normally insurable by special policies such as employer's liability, motor and professional negligence, as well as risks associated with war and radioactive contamination, which are risks for which the government is responsible and are uninsurable. Liability for loss or damage caused to property upon which the contractor is working that are considered inevitable due to the nature of work undertaken is also a normal exclusion. A point to note when public liability policies are taken is that subsidence affecting adjoining properties may not be covered unless a request for its inclusion is made to the insurer; this is a major risk for which cover should be obtained.

Employer's liability insurance

Under clause 24 of the ICE Conditions of Contract the employer is not liable for damages or compensation payable in law in respect of accidents or injury to workmen employed by the contractor or his sub-contractors, except to the extent that they are caused or contributed to by an act or default of the employer, his agents or servants. The contractor is required to indemnify the employer against any claims arising from such accidents and injuries.

There is no corresponding clause in the ICE Conditions of Contract requiring the contractor to insure against this risk because this type of insurance is compulsory in the United Kingdom in terms of the Employer's Liability (Compulsory Insurance) Act 1969. This Act requires an employer to be insured for at least £2 million against this liability.

The employer's liability policies normally have a standard form of wording, no limit of indemnity amount and hardly any exclusions. In order to safeguard against the doubt as to whether labour-only sub-contractors and other forms of gang labour employed on the contract are covered, an endorsement confirming that such forms of labour are insured is made to the policy. When sub-contractors are employed the contractor must ensure that their policies are compatible with the contractor's own policy in regard to cover.

Commentary on insurance clauses

The subject of insurance cover required under civil engineering contracts has been dealt with in the context of the latest (fifth) edition of the ICE Conditions of Contract in the above paragraphs. The principles of insurance underlying civil engineering construction works, whether they be subjected to the ICE, FIDIC, Overseas or a hybrid conditions of contract, could be said to be broadly similar. The general pattern that is followed in conditions of contract applicable to civil engineering construction works is for a clause imposing a responsibility or liability on the contractor followed by an insurance clause relating to such responsibility or liability. As far as excepted risks are concerned on works insurance there are certain differences (cf. Chapter 6) between the ICE Conditions of Contract and those contained in the Overseas and FIDIC. These changes have been made generally to suit the particular situations that are likely to prevail when works are carried out overseas. Similarly, in the ICE Conditions of Contract there is no insurance clause following the clause imposing the legal liability to indemnify the employer against accident or injury to workmen. As previously pointed out, the insurance requirement is not stipulated because such insurance is legally compulsory for any employer in the United Kingdom. In the FIDIC, Overseas and other conditions of contract adopted for overseas construction contracts the insurance requirement against accidents and injuries to workmen is stipulated.

A general point that needs mention is that the insurance clauses in the usual conditions of contract do not recognize the existence of excesses and franchises which are invariably a feature of insurance policies. An excess is the sum of money a policy-holder must bear in the event of a claim made by him. If the excess for a claim is £1000 then the policy-holder would receive £2000 on a £3000 claim. A franchise sum in a policy is the sum above which only claims will be considered. If the franchise sum is £1000 a claim of £1001 will be met in full, whereas a claim for £999 will not be allowed.

The insurance clauses generally require the contractor to take out insurance to the full value of the works. Variation orders, escalation and other factors may increase the contract value during the progress of works, and therefore the adequacy of the sum insured on the policy should always be kept under review.

Placing of construction insurance

Under the terms of the conditions of contract normally incorporated into civil engineering contracts the taking out of three classes of insurance, namely, contractor's all risk, public liability and employer's liability, is imposed on the contractor. Usually the works insurance or the contractor's all risk policy has to be taken out in the joint names of the employer and the contractor. The public liability policy has to be taken out by the contractor on terms approved by the employer. In respect of employer's liability insurance, however, the terms of the insurance cover need not be approved by the employer, although where such insurances are required to be taken out for work overseas a stipulation is generally made in the contract that the insurer shall be approved by the employer and, when required, the policy and receipt of the current premium shall be produced for inspection by the engineer or his representative.

In practice most contractors in the United Kingdom carry annually renewable blanket insurance policies to cover the works undertaken by them at different sites. The scope of such policies is wider than the minimum required under, say, the ICE Conditions of Contract. For instance, in the contractor's all risk policy, professional fees that may be incurred for reinstatement of the works following damage is normally included, although this is not a specific requirement for insurance under the usual conditions of contract. In view of risks of this nature that are inherent in the construction industry it is always a safeguard for a prospective insured to have the liabilities and responsibilities attached to insurance defined by a solicitor and to seek advice from a professional broker, both well versed in construction insurance, before insurances are placed to cover construction risks.

Construction insurance policies for larger civil engineering construction contracts are usually arranged by brokers on behalf of the prospective insured. The broker tends to arrange insurance for a single project with different companies or syndicates that specialize in the different classes of risks involved on a project. On large contracts there can be more than a dozen insurers taking on the risks attached to the differing classes of work involved; such insurance is called co-insurance. Legally, each of the insurers has a contract with the insured, although in practice a leading insurer who is authorized to sign on behalf of all insurers conducts all correspondence relating to the one policy which is issued. The other manner in which insurance is placed is with one insurer who may spread the various categories of risk by re-insuring with companies or syndicates specializing in a particular class of risk.

With re-insurance there is only one contract between the insurer and the insured and the insured is not a party to reinsurance.

It is becoming increasingly the practice of large organizations to arrange a single coordinated insurance programme on complex projects undertaken by them: this type of insurance is also referred to as omnibus, wrap-up, umbrella or total project insurance. The desirability for arranging insurance on a total project basis varies in direct proportion to the size, complexity and the number of parties involved on the project from commencement through to its completion. This type of insurance is most suited on multi-contract, multi-disciplinary projects where several contractors, sub-contractors and others are employed to provide goods and services. An example of such a project would be that for the establishment of a mass transportation system, where several civil engineering contracts which are let on a geographical basis for the construction of stations, tunnels and elevated structures have to be coordinated with a number of other contracts let on a system-wide basis for the permanent way, trains, signalling, power supply, equipment and the like.

Some advantages cited in favour of arranging insurance cover by the employer on multi-contract multi-disciplinary projects are set out below:

1. Wide protection will be available against all fortuitous risks, regardless of the division of responsibility between the separate parties to the projects.
2. All contractors, sub-contractors and others employed to provide goods and services can be jointly covered.
3. Interface problems will be avoided.
4. Cover will not cease on the issue of individual completion certificates but will continue through a project completion.
5. A simple changeover from construction policies to permanent operating insurances will be possible.
6. Disputes about responsibility for damage caused by a combination of factors will be avoided.
7. Third party claims can be settled without dispute as to which party and which insurer is liable.
8. A single comprehensive claims-handling system will be available.
9. Economies of scale and the employer's greater purchasing power will lead to significant premium saving.
10. The need to check individual contractor's policies will be avoided.

Although all insurances required on a project can be brought within the ambit of a single coordinated insurance programme effected by the employer, some insurances, such as employer's liability (i.e. contractor's employees), constructional plant and equipment, and marine cargo, are often left to be insured by the particular contractor. This is because the premiums payable and the degree of cover required for such insurances are dictated and influenced by the standing of the particular contractor, his requirements and his obligations.

Professional indemnity insurance

Under the conditions of contract normally adopted for works of civil engineering construction the contractor is not responsible financially to repair or make good a damage caused to the works arising from a defect, error or omission in its design: that is, unless the design is provided by the contractor pursuant to an obligation imposed on him under the terms of the contract. The responsibility for the design and supervision of the works falls on the party responsible to the employer for the design of the works. Usually this responsibility is exercised by an independent consulting engineer appointed by the employer. The engineer so appointed has certain obligations towards the employer, among which is the exercise of all reasonable skill, care and diligence in the services agreed to be performed by him (cf. Chapter 11).

Should it be proved that a structure designed by the engineer has collapsed due to his negligence in exercising reasonable skill and care, then the liability for damage falls on him. The professional indemnity insurance is a means by which protection can be obtained against the engineer's legal liability to pay damages to third parties who have sustained some injury or damage due to his negligence or that of his staff in the professional practice.

The market available for professional indemnity insurance is small and the cost of such insurance is relatively high. A limit of indemnity is placed on the policy which applies to all claims in any one policy year. These insurances are considered by the insurer on an individual basis and the premium is based on the engineer's previous record, standing, professional expertise and the responsibilities that are taken on by him.

A claim under a professional indemnity policy should be notified to the insurer when there is a suggestion of a claim. Also for a claim to be considered under such a policy it has to be made against the insured during the period of insurance, irrespective of when the cause giving rise to the claim had taken place.

Chapter 10

Tender procedure

Contracts for the execution of projects of a civil engineering nature are normally awarded on the basis of obtaining competitive bids or tenders. The procedure adopted by the employer and the engineer for obtaining such tenders, the evaluation thereof, through to the engineer's recommendation for the selection of a tenderer to whom a contract can be awarded is dealt with in this chapter.

Pre-tender stage

When the type of contract to be adopted for the execution of the project is decided upon, it is for the engineer to prepare tender documents and drawings on which to obtain competitive bids from contractors.

At this pre-tender stage a decision has to be made on exactly who may be permitted to tender. First we have the procedure known as open tendering, where the advent of a tender issue is advertised in the press and/or in technical journals; any person or organization who responds to such a notification is given the documents on which to tender. For most clients this procedure is outmoded, although it remains in certain types of work carried out by local authorities in the United Kingdom and in similar organizations abroad.

The idea of open tendering has developed into a more sophisticated form in respect of large projects, especially where works have to be carried out overseas. For these tenders, prospective contractors are selected after assessing their qualifications to undertake the work envisaged on the particular project. The manner in which this selection is done is termed in the industry as the 'pre-qualification' process. The selection of suitable tenderers by this process is achieved by inviting applications from contractors who wish to pre-qualify for the issue of tender documents by means of advertisement in the press, technical journals or by direct notification to parties who may be interested. The pre-qualification notice calls for evidence from prospective tenderers to demonstrate their technical, management and financial ability to execute the work for which tenders are to be invited.

A pre-qualification notice may require prospective tenderers to provide data relating to matters such as those listed below:

1. Legal status of the company.
2. Particulars of holding, subsidiary or associated companies.
3. The latest published audited annual report and accounts of the company and those in respect of the two previous years.
4. Annual turnover of the company in relation to civil engineering construction works.
5. Statement of construction projects of a similar nature executed by the company together with their values.
6. Number of persons employed in the company's head office and branch offices, excluding those employed in site offices.
7. Names of directors.
8. Numbers employed in management and administrative grades.
9. Numbers employed in design and drawing offices of the company.
10. Construction plant and labour resources of the company.
11. References, membership of trade associations, permission to visit some of the works the company had executed and other relevant data.

Usually a time limit of between two and six weeks is set for the submission of the foregoing data. Contracting organizations are generally used to preparing and presenting this information, and at times the documents submitted by one party can run into several volumes. Depending on the project and the market situation, when international contractors are sought for the execution of a project the number of organizations who seek pre-qualification can be as many as 50. On receiving the pre-qualification data the engineer assesses them and recommends to the employer those organizations that are in his opinion suitable for the issue of tender documents for the project under consideration. If there is a large number of suitable applicants it may be necessary to limit the number from whom tenders are to be invited, preferably to not more than eight. A very large spread of tenderers has the effect of dissuading otherwise purposeful organizations from submitting competitive bids. In this context it is a fact that the preparation of a tender for a major civil engineering project is very expensive with a low chance of producing any return. The limitation of prospective tenderers to a specific number, although desirous, is not always possible. For instance, on projects financed by the World Bank and certain other institutions all who meet the pre-qualification criteria must be permitted to tender. When the engineer's recommendation is agreed by the employer tenders are invited from the contracting organizations thus selected.

The second method of obtaining competitive bids is to invite tenders from selected contracting organizations prepared by the engineer in conjunction with the employer. Lists of contracting organizations suitable for various civil engineering projects are compiled by both the large promoting organizations, and consulting engineers compile such lists from either first-hand knowledge of the contractors or by assessing the capabilities and the capacity of contractors who are known to execute projects of a civil engineering nature. Only contracting organizations of repute with adequate financial and technical resources for the various classes and scales of projects are included in these lists. The lists are reviewed periodically so that new firms of good standing can be included and those whose performance is subsequently known to be wanting can be deleted. The invitation of tenders from contracting organizations selected from lists prepared by the engineer in conjunction with the employer is the most usual method nowadays in the United Kingdom. It is also generally considered to be the best method in that meaningful tenders can be obtained with the least delay.

Lastly, a contract may be negotiated with a selected contractor, and this procedure is known, by definition, as a single-tender nomination. This procedure should only be adopted in special cases, such as the carrying out of an extension of a project which has already been completed by a particular contractor, and it may be done on the existing rates or as a percentage addition or reduction to existing rates or, of course, on some form of cost/reimbursement basis. The procedure is also used where some unique or special expertise is required.

All the above methods are used for the selection of a contractor competent to carry out works envisaged in the contract. The size of the job, its location and its complexity are the normal criteria that dictate the adoption of a particular method of selection. However, personal preference, tradition and local custom are among matters which influence the manner of calling tenders.

About one week before the tender documents are sent out for enquiry it is good practice for the engineer to issue to the employer his estimate of the tender price. It is unfortunately true to add, however, that unless an employer specifically asks for it, the engineer has been known to be guilty of not carrying out this duty, purely because he may be afraid that his estimate will bear little relationship to the lowest tender received. Nonetheless it is a good discipline, and it is a practice which should be adhered to by the competent engineer.

We come next to the tender period, and it is wise for this to be of reasonable length in order to allow tenderers time to obtain bids from their enquiries as far as materials and specialized equipment are concerned and to study and choose the optimum methods to execute the works. In the United Kingdom one usually thinks in terms of six weeks for major works, although, here again, the engineer is often subject to pressure by the employer to reduce the tender period. In the case of international contracts a three-month tender period is usual, the minimum being two months. The disadvantage of a short tender period is that the tenderers will not have the time to plan the work properly and economically and price the work in a realistic manner. When this happens tenderers invariably qualify their tenders and the apparent time saved is then lost during the tender-adjudication period and also on inevitable delays that can arise during the construction period as a result of inadequate planning of work before the tenders are submitted.

Tender stage

The tender stage can be said to commence when the tender documents are despatched to the selected tenderers under the cover of a letter of enquiry. This letter is normally short because the instructions to tenderers bound into the tender documents set out the manner in which the tenders are to be prepared, stipulate the closing date and tell how the tenders are to be delivered. The letter may highlight some of these requirements but must call for the documents to be acknowledged as soon as they are received by the tenderer.

On major contracts the necessity for revisions to the tender documents may come to light during the tender stage. Such revisions are communicated to all tenderers by issue of tender memoranda. These memoranda are serially numbered and bear the caption that 'they are deemed to be an integral part of the contract documents and drawings'.

Provision is normally made in the 'instructions to tenderers' for queries on the documents to be addressed to the engineer for clarification during the tender stage. Such queries fall into two categories. The first is where the question raised can be answered without any ambiguity by reference to the documents already issued to tenderers or where it may be that the particular query cannot be answered. In such cases the tenderer may be referred to the relevant clause in the document that answers the question or when the query cannot be answered he should be so told; other tenderers need not be informed of queries of this nature. The second category consists of queries relating to an ambiguity in the documents, an error or additional information which is required and should have been provided in the documents. In cases such as this the queries must be answered and the answer given will have a bearing on the other tenderers aspiring to win the contract. A practical manner in which to deal with such questions is to embody the clarification to the question in a tender memorandum and issue this to all tenderers.

A procedure which merits consideration and is sometimes adopted for clarifying matters connected with the tender is for the engineer to hold a site meeting with all tenderers present. At such a meeting an opportunity is afforded to all tenderers to raise any queries, especially in relation to the site, which are then clarified and put together in a memorandum which eventually forms a part of the contract.

Tenderers for major works often find that the closing date of tenders is difficult to meet. In this event they usually ask for an extension to the date for submission of tenders and reasons are given for such a request. The engineer, in consultation with the employer, has to consider whether these requests are reasonable. Again in practical terms, such extensions should not be given if a request is received late in the tender period such that other tenderers would be at a disadvantage or if only one tenderer out of six asks for an extension. On the other hand, if four out of six tenderers require an extension it would be unwise for the employer not to accede to such a request.

The tenderers are required to lodge the tenders by the due time and date set out in the 'instructions to tenderers'. Should the tenders be late in arriving they are liable to be rejected. Postal or transport delays could occur resulting in tenders ariving late, although they have been despatched in good time. To circumvent this possibility tenderers often take the additional precaution of informing the appropriate organization by telex or facsimile of the tender total and the date and manner of despatch of the documents. Such advice, when received by the due time on the date set for receiving tenders, is accepted by some authorities as adequate proof of despatch and the tender documents which arrive shortly afterwards are then considered.

The instructions to tenderers state explicitly the place where tenders are to be lodged. These could be with the employer or with the engineer acting on behalf of the employer. Often the requirement is for the original of the tender documents to be submitted to the employer and a copy thereof to be sent to the engineer.

Utmost propriety should be exercised by all concerned in handling tenders. Most large promoting organizations have formal procedures laid down for this purpose. Some matters that are dealt with in a formal procedure may be stated to be as follows:

1. On receipt of tender packets, they are endorsed with the time and date of receipt.
2. The un-opened packets are then placed in a tender box. This may be especially constructed for safe-keeping of tender packets or may be a safe, cupboard, strong room or other accommodation which is secure and lockable. The custody of the keys to such accommodation will be with an authorized person.
3. A tender board or a tender panel is appointed by the promotion organization or the consulting engineer, as the case may be. Such a board or panel is authorized to deal with the opening of tenders and the formalities that flow from it. The constitution of this board depends on the organization concerned; it is normal for at least three members to be appointed to form such a board with the chairman being a principal or a senior executive of the organization.
4. At the due time set for opening the tenders, which could be a time immediately after that fixed for receipt of tenders or a later time or date, the tenders are opened by the board. Just before opening, all packets are counted and a serial number is attached to each packet. When the packets are opened the tender letter, the completed tender form and the general summary of the bill of quantities (where such are enclosed) are initialled by each member of the board. The prices of the tenders received are then recorded in a form designed for this purpose and signed by the members of the board. In certain cases the opening of the tenders is in public or restricted to the tenderers. When this practice is adopted the tender totals are read out so that those present can note the prices of all parties that have tendered for the works. In giving this information the point is always made that the prices read out are the face values of the tenders as received, and that they are liable to change when the tenders are evaluated.

Post-tender stage

The first step that should be taken after the tenders are opened is for the engineer or the employer to acknowledge the receipt of the tenders.

The tenders received are evaluated by the engineer. If the documents have been received by the employer and no copies are submitted to the engineer either a copy or the original of all documents received is made available to the engineer.

During the process of evaluation strict confidentiality is exercised by the engineer and the team charged with this responsibility on all matters connected with the documents submitted by tenderers.

The purpose of a tender evaluation is to determine the cost to the employer of each tender in comparison with others. The lowest submitted tender price may not be the lowest evaluated cost. During the evaluation process the details of the tenders submitted are examined to ensure that they conform with the tender documents. They are arithmetically checked and other relevant factors such as the qualifications imposed by the tenderers, programme requirements, deviations from specification, mobilization periods for staff and equipment and special terms of payment are considered. Where possible, any deviations from the requirements of the tender documents are expressed in monetary terms so as to obtain the comparable cost of the tenders.

The detailed manner in which tenders are evaluated depends on the type of contract for which tenders have been called. For instance, when lump-sum, cost/reimbursement or turnkey contracts are under evaluation the details which require examination naturally differ from those which would be analysed in respect of an admeasurement contract. Again, the detailed manner in which an admeasurement contract in itself would be analysed depends on the scale and type of the works envisaged and also to no less an extent on the practice adopted by different engineers. As an example, some of the matters that are examined in respect of an admeasurement contract for a major civil engineering project with a relatively minor mechanical and electrical content may be expressed in the following terms:

1. *Verification that all documents called for in the tender documents have been submitted*
   All the documents received from a tenderer should first be listed. These include the tender letter, the documents called for in the instructions to tenderers such as outline program-

mes, phased labour charts, tender bond (where applicable), names of proposed sub-contractors, list of plant to be used on the works and, of course, the completed tender documents.

2. *Legal status of the contractor*

This aspect normally only needs to be examined at the tender-evaluation stage if this has not been evaluated prior to the issuance of tender documents. However, a situation may arise on international contracts where joint ventures of two or more contractors bid for a contract. In such cases a copy of the pre-bidding joint venture agreement which had been entered into to bid for the contract would be enclosed if required together with the draft terms of a joint venture agreement which would be entered into if the contract is awarded to such a joint venture. The terms of the documents thus enclosed will have to be examined by the assessors to ensure that they are in order. Another point to note is that when dealing with a joint venture or a subsidiary company it may well be necessary to obtain a guarantee from the respective holding companies in respect of the work undertaken by them.

3. *Prices and terms of payment*

Prices and terms of payment are stipulated in the tender documents, and accordingly there should be no argument in this respect in theory. In practice, especially on international contracts, tenderers invariably require some amendment to the stipulations made in the tender documents. These include matters such as payments against letters of credit, advance payments, mobilization costs payable in advance, payment for specialized plant on delivery to site, payment to be effected in foreign currency at a fixed rate of exchange, exemption from import duties and other variants which all have an effect on the ultimate cost of the works to the employer. During evaluation it is necessary to assess the effect, if any, of such conditions as are imposed and to determine their monetary values in order to bring all tender totals to a common comparable base.

4 *Special conditions imposed by tenderers*

Although the conditions of contract that are to be adopted are enclosed with the tender documents, qualifications are sometimes imposed on them by tenderers, once again mainly on international contracts. For instance, a tenderer may require energy shortages or energy-related shortages in the country where the construction is to take place to qualify for an adjustment of the tender price. Likewise, he may require delays over and above a specified period arising out of port congestion automatically to qualify for an extension of time and delay costs. It may even be a requirement to accept a retention bond in lieu of retention monies required to be withheld under the relevant clause in the conditions of contract. In evaluating these qualifications it is the assessors' task to place monetary values against these qualifications, if that be possible. These values should then be added to or deducted from the tender total to arrive at the evaluated comparable tender total. Notwithstanding having adjusted the tender total, the tenderer could be asked to withdraw such a condition if such is not acceptable.

In respect of qualifications which have no financial significance and are of a minor nature the tenderers could be asked to withdraw them. If the financial significance of a qualification cannot be quantified because the risk attached to such qualification cannot be assessed at the time of evaluation then the contractor may be asked to withdraw it, and if he does so this should be noted. It should always be borne in mind that, other matters being equal, the tenderer who has not made a qualification to the tender should in fairness be recommended for acceptance.

5. *Programme of works*

Under the Institution of Civil Engineers Conditions of Contract or under most standard conditions of contract a construction programme need not be called for at the time of submitting a tender: the requirement under such conditions of contract being to stipulate the time by which the whole of the works are to be completed or the periods of completion of sections of the works. The lack of any programme of works with the bid is not satisfactory. The engineer is advised to call for such a programme as part of the tender submission to evidence the tenderer's appreciation of the scope of the works during the adjudication of tenders.

6. *Errors and omissions*

The first check that is carried out on a tender is an arithmetical one. On an admeasurement contract the extension of the rates and quantities as well as the item totals are checked.

Should there be any errors in the extensions, the rates are considered binding unless there are genuine mistakes in them; a possible genuine error would be when a tenderer enters a rate of £50 per cubic metre for concrete against, say, ten items and £5 per cubic metre for a few others. All arithmetical and other errors in the tender should be brought to the notice of the tenderer and his agreement obtained for the corrections.

A tenderer may omit to insert a price against an item in the bill of quantities by inserting the words 'not included' against such an item. Strictly, such a tender could be rejected on the grounds that it is incomplete. On the other hand, there may be a genuine reason for the tenderer not being able to price such an item; for instance, the only source from which that item could be obtained had not given a quotation or it may be for another genuine reason. In cases such as these the engineer may have to provide a provisional sum to cover the item when evaluating the tender.

7. *Examination of rates*

The first exercise that is carried out for comparison of rates is for a vidimus or conspectus to be prepared. This is a table listing all items in the bill of quantities against which the rates and prices quoted by each tenderer are entered together with the engineer's own estimated rates and prices for these items. Ideally, the first column of figures should be that of the engineer's estimate followed by that of the lowest tenderer, proceeding in an ascending order with the last column containing the figures in the highest tender. A red line should be drawn under each tenderer's item which is the lowest in terms of the unit rate. A proper study of the vidimus will show whether any unit rate quoted by any tenderer is inconsistent or whether such a rate is unduly weighted. The vidimus will also highlight any obvious errors in the rates such as, say, £5 per cubic metre of concrete being inserted in lieu of £50. In effect, the vidimus is a useful tool in the hands of the engineer for the examination of rates and prices quoted by tenderers.

After examining the tenderer's unit rates and prices it is necessary for the engineer to assess whether any of the bids are unbalanced. This means some rates *per se* are unrealistic and others are loaded. It cannot be entirely ruled out that a tenderer purposely does this to ease his cash flow problems. When a bid is unbalanced particular care should be taken to re-check the quantities and also bring this matter to the particular attention of the employer in the adjudication report. Depending on the degree of inbalance, such a tender may have to be rejected.

In a civil engineering contract the major cost of construction generally falls into a few items such as excavation, concrete, formwork and reinforcement. These may represent about 70 per cent of the total cost, say, on a power station contract. Consequently the total cost tendered for these items by each tenderer is normally extracted so that the engineer can assess the pricing patterns of the different tenders on major items of work.

8. *Deviations from civil engineering specification*

The tenderers are required to price their bids in accordance with the specification enclosed with the tender documents. Notwithstanding this requirement, tenderers sometimes offer alternative specifications in respect of some items, especially on major contracts. An alternative specification may be offered because the engineer's specification for a specialized product may be too onerous or the contractor may not have the plant resources to execute a part of the works in the manner specified, or it may be that he considers that a particular item of work cannot be carried out in the manner specified. Indeed, many specifications drawn up by the engineer allow for minor alterations to be offered to works designed and specified in the documents, with the proviso that in such cases a tender should be submitted in accordance with the documents in addition to one based on proposed alternatives. In certain instances the tenderers will only price the alternatives. Strictly, non-compliance with a requirement in the specification makes such a tender liable to rejection. However, when the changes proposed are of a minor nature and the reasons adduced for such changes are genuine it may be in the interest of the employer to consider them. In the evaluation of tenders it would then be necessary for the engineer to estimate the prices that the tenderer should have quoted for such items according to the engineer's specification to arrive at the evaluated comparable tender total. The practice in such instances is for the engineer to insert the highest rate quoted by any of the tenderers for such an item conforming with the engineer's specification unless, of course, such a rate is completely unrealistic.

9. *Deviations from the mechanical and electrical specification*

In the tender documents prepared for a civil engineering project with an element of mechanical and electrical work content, it is evident that it is necessary to include a specification devoted to the mechanical and electrical work. This work may include small power and lighting, heating and ventilation, air conditioning, workshop equipment and mechanical handling equipment. In addition to the specification a schedule of technical requirements and a schedule of technical particulars would be included with the tender documents. The former schedule lists the engineer's requirements in respect of plant and equipment and the latter requires the contractor to list the particulars in respect of the plant and equipment offered. The rates and prices inserted into the bill of quantities are to be for plant and equipment which comply with the specification and the requirements listed in the relevant schedule.

In evaluating this aspect of the offer the engineer has to satisfy himself that the plant offered and listed in the schedule of technical particulars is consistent with the specification and the requirements listed in the engineer's schedule. Any deviations therefrom should be discussed and clarified with the tenderers before a recommendation on the award of the contract is made to the employer. The engineer may in such cases have to take into account any financial effect of items that do not conform to the specification in arriving at the evaluated comparable tender total.

10. *Mobilization of staff and labour*

On larger projects, especially overseas, it is necessary during the tender-evaluation stage to ascertain the tenderer's intentions regarding personnel he proposes to employ on the works. Although it is clearly the responsibility of the contractor to provide the necessary manpower resources for the execution of the works efficiently and complete the contract within the stipulated period, this aspect has to be examined in some detail when expatriate management and technical personnel and imported labour are envisaged to be employed. Again, the mobilization period for personnel has to be ascertained in the context of housing, recruitment and immigration formalities that have to be adhered to in many countries.

11. *Plant and equipment*

The deployment of plant and equipment required for the execution of the works is again a clear responsibility of the contractor. However, at the tender-evaluation stage it is necessary to ascertain the major items of plant that the tenderer proposes to use and assess whether they are generally adequate for the execution of the works. Also the condition of the plant (i.e. whether it is new or used), the delivery periods, shipping times and other relevant data should be ascertained from the tenderers. The dates of mobilization of the plant at site should also be elicited and clarified.

Initial assessment of tenders

On making an initial assessment of tenders it becomes apparent to the engineer that some of the tenders received are no longer in contention. This can be on the basis of high prices or due to their not conforming to the requirements stipulated in the tender documents. In respect of the balance it becomes necessary for certain matters to be clarified and confirmation thereof obtained from the tenderers. In order to obtain such clarification it is normal for the engineer to hold separate meetings with each such tenderer. The number of tenderers with whom meetings are required vary; on a major civil engineering contract this could at times be as many as five. At these meetings the tenderer is informed at the outset that holding a meeting does not in any way imply that the engineer's recommendation will be in their favour. All matters required to be clarified are raised at such meetings and clarified. Minutes of the meetings are recorded and these become an integral part of the tender in the event of a contract being awarded to the party concerned. In addition to the meetings matters are also clarified by correspondence, which eventually also forms part of the contract.

Tender-adjudication report

The engineer, having examined the tenders and evaluated them, is required to present an adjudication report to the employer. The engineer's findings on the tenders received are incorporated into such a report. Adjudication reports run into several volumes on some major contracts and all this work has to be carried out correctly and expeditiously: the normal validity period stipulated for major tenders is 90 days, i.e. the period from receiving the tenders to the award of a contract. At times this period is extended with the concurrence of the tenderers but the fact remains that normally an adjudication report should be presented to the employer within six to eight weeks of receiving a tender, leaving approximately six to eight weeks for the employer to go through his formalities and award a contract. It follows that an adjudication report should be thorough and all matters on the tenders received should be properly evaluated and incorporated into the report. The headings of the subject matter covered in an adjudication report may be set out as follows:

Introduction
Recommendation

Particulars of tenders received
Initial elimination of tenders
Comparison of tenders
Conclusion

Along with the recommendation a draft letter of acceptance may be included. The conclusion should set out a technical assessment as well as a financial assessment of all tenders received. The report will also contain a series of appendices in the form of tables showing the details of the tender figures, tender totals, evaluated comparable tender totals, records of minutes of meetings, copies of relevant correspondence with tenderers after the tenders were received and other data, as appropriate.

Chapter 11

Powers, duties and functions of the engineer and his representative

The engineer is the person or persons duly appointed by the promoter to take on the overall engineering responsibility for the establishment of a project of a civil engineering nature. The duties of the engineer include initial studies, feasibility reports, design, preparation of tender and contract documents and drawings and the supervision of construction during the execution of the project.

Qualifications

The engineer is normally a corporate member of a recognized technical institution and subject to its rules of professional conduct. Often, in the United Kingdom, the engineer is a consulting engineer who is also a member of his Association of Consulting Engineers, in which case he will be subject to their professional rules and conduct. Alternatively, the engineer may be an employee of the promoter.

The consulting engineer has developed over a long period of time into what may arguably be called an elitist section of the engineering profession. By far the greatest impact on the profession has been in the United Kingdom, where, until the beginning of this century the term 'consulting engineer' was not defined and, indeed, consulting engineering work and contracting were to no small extent interwoven.

In the United Kingdom it was in 1908 that the Association of Consulting Engineers was formed, and this lays down the aims, objects, functions and duties of consulting engineers. This body of independent qualified engineers carried out, *inter alia*, duties of the engineer in all parts of the world, and those with experience in this particular field argue with validity that the general United Kingdom system of having an independent technical administrator is both the fairest and best method of carrying out major engineering works.

Appointment

The engineer is best chosen by the promoter on the record of his previous experience and performance on projects of a nature similar to that envisaged. In the United Kingdom an appointement of a consulting engineer is made by a promoter either directly or through fee competition. The ability of the consulting engineer to render the required services is judged by the promoter from his personal knowledge, a personal recommendation, by reference to the Association of Consulting Engineers or by inviting prospective consulting engineers to pre-qualify for submission of proposals through a notification in the technical press. Select lists of consulting engineers with their respective fields of specialization are maintained by most government departments, local authorities and public companies. On overseas projects the appointment of consulting engineers is made much in a manner similar to that obtaining in the United Kingdom. Lists of consulting engineers with their experience are available with trade commissioners or commercial attachés of embassies abroad as well as with international financing institutions such as the World Bank, the Asian Development Bank and similar organizations.

It is unfortunately true to state that fee competition has become more common in recent years. While competition in technical competence is not unreasonable, the lowest offer in terms of money leads all too often to an inferior service and with it a capital project cost far in excess of any increase in consulting engineer's fees: a lower-quality article is a corollary.

Agreement between promoter and engineer

When the promoter has selected a consulting engineer for appointment as the engineer for a project an agreement is entered into where the obligations of the parties are set out. Such an agreement is in writing and, if required, executed under seal. The Association of Consulting Engineers' model forms of agreement are widely used in the United Kingdom for consulting engineering services. These model forms, as well as those published by the Fédération Internationale des Ingénieurs-Conseils (FIDIC), are also used or serve as a guide when entering into service agreements with promoters on overseas projects. On many projects, the engineer's services are engaged only by exchange of letters; in such cases it is usual for reference to be made to the conditions of engagement contained in a model form such as published by the Association of Consulting Engineers or a similar body.

Some of the matters covered in an agreement for an engineer's professional services on a project of a civil engineering nature may be expressed in the following terms:

1. Duration of engagement.
2. Ownership of documents and copyright.
3. Arbitration clause for settlement of disputes which may arise out of the engagement.
4. Scope of consulting engineer's services.
5. Information and services (if any) to be provided by the promoter to the engineer.
6. Terms of payment.
7. Fees payable in the event of postponement, cancellation or abandonment of the project.
8. Effect of *force majeure* on the services to be rendered by the engineer.
9. The responsibilities of the parties in respect of taxation, customs duties and other dues.
10. Places at which notices are to be served under the agreement
11. Language to be used in correspondence and documents in connection with the agreement.
12. Law applicable to the agreement.

Certain points that require to be noted when entering into a service agreement are that:

1. The engineer must ensure that his appointment is within the scope of the promoter's authority and
2. The engineer's authority is terminated on bankruptcy of the promoter.

Legal position of the engineer

The legal position of the engineer is governed by a personal contract between the promoter and the engineer. In the context of the construction contract entered into between the promoter (i.e. employer in terms of the definition of the promoter given in standard forms of contract such as issued by ICE and FIDIC) and a contractor, the engineer acts as the agent of the promoter. The engineer can be sued by the promoter for negligence, incompetence and for not exercising reasonable care and skill in respect of his duties.

Feasibility study stage

On the appointment of an engineer the first duty that is required to be performed by him is usually the preparation of an engineering feasibility study for the project.

The feasibility study forms a fundamental prerequisite to committing expenditure on a project, for which the engineer is responsible. It is essential that the engineer should know the promoter's intention regarding the terms and programme of capital expenditure.

The promoter should provide the engineer with all the information he has available and explain fully his requirements. Unless mutual understanding and confidence exists between the promoter and his engineer the latter's report may not be fully effective.

Once the promoter's intention is ascertained, the engineer will have to establish, in agreement with the promoter, the essential requirements of the project, eliminating at this stage all features which will not unduly influence the main technical and economic considerations. Having acquainted himself of these basic requirements, the engineer will then generally undertake investigations to enable him to apply his experience and formulate reliable recommendations to the promoter.

The results of the engineer's investigations will be presented to the promoter in the form of a report. The style and form of this are at the discretion of the engineer and will vary according to the nature of the project. It is necessary that the subject matter of the report is so presented that it can be readily understood by the promoter, who may neither be able to appreciate the finer technical details nor have the time to absorb them. On the other hand, the report must contain sufficient technical information to convince any other suitably qualified engineer to whom it may be referred of its soundness. It is often found convenient to include subject matter in appendices to the main report or in a separate volume.

The report generally contains the following aspects as appropriate:

1. Terms of reference.
2. Available data.
3. Basic assumptions, data or trend lines on which the engineer's solution is based, and his reasons for adopting these.
4. Details of investigations carried out and design criteria adopted.

5. Alternative sites where appropriate and solutions considered with reasons for their rejection.
6. Scheme proposed with alternatives where appropriate.
7. Geological and geotechnical aspects.
8. Capital costs.
9. Operating costs.
10. Economic value.
11. Allocation of costs if chargeable to several organizations.
12. Labour and principal materials requirement.
13. Master programme, including the period of construction.
14. Summary of conclusions.
15. Outline drawings.

Design and construction stages

When the promoter makes a decision to go ahead with a project on the basis of the engineering feasibility report or otherwise he then appoints an engineer to take on the overall engineering responsibility for the design, preparation of tender and contract documents and the supervision of construction, completion and maintenance of the project. Often the engineer who prepared the engineering feasibility study is then instructed to discharge this function, but this is not always the case. The duties and responsibilities of the engineer in discharging this function could best be dealt with by describing his general duties and responsibilities during the various stages through which the project is realized: these can be classified broadly as the preliminary design; design and tender; and the construction.

Preliminary design

During the preliminary design stage the engineer has to define the engineering work involved in such a manner that the promoter can be apprised of the work involved in the project and the means by which this can generally be executed. The engineer is responsible at this time for determining any easements or restrictions at site and to satisfy himself regarding all facts relating to site conditions. The work carried out by the engineer at this stage will include preparation of outline drawings, estimates and other documents to obtain the promoter's approval in principle for the works and to enable him to obtain planning permission for the works from the appropriate government department, public authority or such other body. The other duties performed by the engineer where relevant, during the preliminary design stage may be expressed in the following terms:

1. Preparation of topographical survey or surveys at the site.
2. Investigation and examination of soil conditions at site.
3. Investigation of available data and collation of factual information relating to the works.
4. Advising the promoter on any investigations necessary to be carried out such as sinking of trial boreholes, field investigations, plate-bearing tests, hydrographical surveys, laboratory tests, models and other relevant investigations or tests.
5. Consultation with any architect or services engineer appointed by the promoter in regard to architectural treatment and mechanical and electrical services, respectively.
6. Making modifications as appropriate on the outline drawings and estimates for the works resulting from the requirements of the architect or the services engineer as may be approved by the promoter.
7. Updating the estimates of capital cost and master programme.

Design and tender

The engineer's function during the design and tender stage is to define the works in a manner so that they can be executed by a contractor. This function is discharged by first preparing the designs, tender drawings, specifications and the other documents which are required for obtaining competitive tenders leading to the appointment of a contractor to execute the works. There are times when competitive tenders may not be required because the employer so wishes, or for some good practical reason it is considered that the works should be carried out on a single-tender nomination or directly through the employer's own or hired labour. At all events, the scope of the work to be executed has to be properly defined at this stage. The elements of work performed by the engineer during this stage may be summarized as follows:

1. Preparation of design and tender drawings for the works.
2. Assessment of materials and manufactured articles that are necessary for the works.
3. Preparation of specifications, bills of quantities, and schedules, as appropriate.
4. Assessment of provisional sums and prime cost items that are required to be incorporated into the bill of quantities.
5. Identification of the type of contract that is to be adopted (i.e. admeasurement, lump sum, or cost/reimbursement, etc.).

6. Preparation of tender documents and drawings for the construction contract.
7. Preparation of tender documents for obtaining nominated sub-contractors, as appropriate.
8. Identification of the manner in which tenders are to be called (i.e. open tenders, selected tenders, or single-tender nomination).
9. Assessing the qualification and suitability of prospective tenderers and advising the promoter on their suitability for submitting tenders.
10. Updating estimates of capital cost and master programme.
11. When so required, inviting tenders on behalf of the promoter from tenderers approved by him.
12. Evaluation of tenders when received, preparation of an adjudication report and recommending to the promoter the tenderer most suitable for carrying out the works. A draft letter of acceptance to the recommended tenderer is often appended to the adjudication report.

Construction

On the basis of the recommendation made by the engineer a letter of acceptance is normally issued by the promoter to the selected tenderer, thus forming a contract between the employer (i.e. the definition for the promoter given in the standard forms of contract) and the contractor, for the due execution of the works by the contractor.

During the construction stage the responsibility falls on the engineer for the performance of a multitude of duties towards the realization of the project. These duties flow from both the service agreement he has entered into with the employer and from those imposed on him by the construction contract entered into between the employer and the contractor. The duties that the engineer has to perform in relation to the construction contract are obligations which, of course, in turn flow from the service agreement between the employer and the engineer. The obligations of the engineer towards the employer during this stage for a project of a civil engineering nature may broadly be expressed as follows:

1. Advising the employer on the preparation of the formal contract documents relating to the accepted tenders.
2. Advising the employer of staff to be employed at site by the engineer (or the employer) for the supervision of the works.
3. Preparation of working drawings, bar-bending schedules and

other designs and details that are required for the due execution of the works.

4. Examining the contractor's proposals.
5. Making such visits to site as the engineer considers necessary to satisfy himself as to the performance of any site staff appointed on the engineer's advice and to satisfy himself that the works are executed generally in accordance with his designs and specifications and otherwise in accordance with good engineering practice.
6. Giving all necessary instructions to the contractor under the construction contract or contracts. Such instructions should not significantly increase the cost of the works without prior approval from the employer, unless the circumstances were such that this was not possible. In the latter event the employer should be advised of any increases in cost as soon as it is practicable. As a matter of good practice the engineer should prepare and issue to the employer a revised estimate of the final capital cost every three months on works of any complexity.
7. Performing any services which the engineer may be required to carry out under any contract for the execution of the works provided the engineer had initially approved such contracts.
8. Providing the employer on completion of the works with such records and drawings as are reasonably necessary to operate and maintain the works in respect of the element of work supervised by the engineer.
9. Assisting the employer in settling disputes that may arise between him and the contractor, excepting litigation and arbitration, which are dealt with under a separate arrangement.
10. All other duties imposed on the engineer by the construction contract entered into between the employer and the contractor.

The engineer's role in the construction contract

When a contract is entered into between the employer and the contractor for the construction of the works a distinct role falls upon the engineer in administering the contract. The engineer's powers and his duties for the administration of the contract flow from the conditions of contract incorporated into the construction contract. In the discharge of his functions under the usual forms of contract such as those issued by the ICE and FIDIC the engineer alone is responsible for any instructions, directions and decisions on any matters related to the contract; both parties to the contract are protected by their right of appeal under the arbitration clause. The engineer is entitled to seek legal advice as to the interpretation of the contract and he may ascertain matters of fact from many sources, but he should not accept instruction by the employer on the attitude he is to adopt in the discharge of his duties. The engineer must be in a position to act in an independent role in administering the contract and exercise his judgement on matters arising therefrom fairly, without showing any bias towards either the employer or the contractor. This aspect is covered by the ICE Guidance Note 2A: 'Functions of the Engineer under the ICE Conditions of Contract.'

The duties and responsibilities conferred on the engineer by the contract are expressed in the conditions contained therein. Perhaps the most fundamental clause in relation to his duty is expressed in clause 13(1) of the ICE Conditions of Contract, which states:

'Save in so far as it is legally or physically impossible the Contractor shall construct complete and maintain the Works in strict accordance with the Contract to the satisfaction of the Engineer and shall comply with and adhere strictly to the Engineer's instructions and directions on any matter connected therewith (whether mentioned in the Contract or not). The Contractor shall take instructions and directions only from the Engineer or (subject to the limitations referred to in Clause 2) from the Engineer's Representative.'

The first task the engineer is required to perform under the contract is to issue an order to commence work. Following this order it is normal for the engineer to invite the contractor to a meeting at his office to clarify and apprise him of the procedure to be adopted for the smooth operation of the contract. Some of the matters that are discussed and established at such a meeting are:

1. *Organization and correspondence*
   The engineer's and contractor's personnel to be associated with the works are established. The names of the engineer's representative, usually called the resident engineer, and the contractor's agent proposed to be employed at site are disclosed. It is usual for the contractor to be asked to confirm in writing the name of his proposed agent with his curriculum vitae and for the engineer to signify his approval also in writing, provided the person proposed is acceptable. The addresses and the manner in which correspondence is to be routed with the number of copies of correspondence required is identified at

this stage. The lines of communication between the engineer's and contractor's organizations are also established. It is normal to stipulate that direct communications will be between the contractor's head office and the engineer's head office, on the one hand, and the contractor's site agent and the resident engineer, on the other, but never between the contractor's site agent and the engineer's head office, or between the contractor's head office and the resident engineer. Likewise the lines of communication should be confined to between the site agent and his head office and between the resident engineer and his head office.

2. *Contract documents and drawings*
   Insurance certificates in respect of insurance policies required to be taken out under the contract and performance bonds, if any, should be called for. The number of copies of contract documents and drawings required by the contractor are to be agreed.

3. *Programme*
   The contractor's detailed programme of works is called for. The date for its submission is to be agreed.

4. The phased labour chart together with a statement on the manner in which constructional plant is to be deployed should be called for.

5. A further meeting for the handing over of the site on the date for commencement of the works should be fixed.

6. Any other matters that are relevant and are peculiar to the contract should be identified. These include the manner of presenting interim certificate measurements, the procedure for rate-fixing, establishment of letters of credit in the event that letters of credit form the basis for payment, the manner in which advance payments, if any, are to be dealt with, the demarcation of sub-contractors' interfaces with other contractors to be employed, and the like.

As stated previously, the engineer has at this stage to perform duties which flow from both the service agreement he has with the employer and from the construction contract. Some of the duties that particularly stem from a usual construction contract adopted for works of a civil engineering nature may be expressed as follows:

1. Supervision of the works.
2. Issue of further drawings and instructions for the due execution of the works.
3. Examining and approving or consenting to contractor's

proposals, construction techniques, temporary works or construction plant.

4. To satisfy himself that the quality of materials to be used on the works conforms to the specification.
5. Inspecting and testing during manufacture of such materials and plant for the permanent works as are usually inspected and tested by the engineer.
6. Supervision of acceptance tests on site.
7. Issue of orders to vary the works.
8. Issue of instructions for committing expenditure under provisional sums and prime cost items in the bill of quantities.
9. Admeasurement of works.
10. Preparation and issue of payment certificates.
11. Rate fixing.
12. Assessment of claims submitted by the contractor.
13. Giving an engineer's decision on any dispute or difference which may arise out of the contract.
14. Issue completion certificates and maintenance certificates.

Engineeer's representative

The function of the engineer's representative is to watch and supervise the construction, completion and maintenance of the works. He shall have no authority to relieve the contractor of any duties or obligations under the contract nor, except as expressly provided in the conditions of contract, to order any work involving delay or any extra payment by the employer nor to make any variation of or in the works.

Delegation by the engineer

The engineer may from time to time, and at the commencement of the contract, authorize the engineer's representative to act on behalf of the engineer generally in respect of the contract or specifically in respect of particular clauses of the conditions of contract within the scope of his authority. Notice of such authorization should be transmitted to the contractor by the engineer in writing. In the ICE Conditions of Contract such authorization is specifically excluded in regard to any decision to be taken or certificate to be issued under the following clauses:

Clause
12(3)   Delay and Extra Cost

44	Extension of Time for Completion
48	Certificate of Completion of Works
60(3)	Final Account
61	Maintenance Certificate
63	Forfeiture
66	Settlement of Disputes – Arbitration.

A typical list of matters delegated to the engineer's representative flowing from the clauses in the conditions of contract may be expressed as follows:

Clause

2	Supervision of workmanship and materials
7	Issue drawings and receive notice of further drawings required
12	Receive notice of unforeseen adverse physical conditions and artificial obstructions
14	Receive programmes
15	Give instructions to contractor's agent
16	Object to and require removal of persons employed by contractor except contractor's agent
17	Require rectification of errors in setting out
18	Require additional boreholes
19	Require lights, guards, fencing and watching
20	Require making good of damage to the works
30(3)	Receive reports of claims in respect of damage by transport of materials or manufactured articles
31(1)	Require facilities for other contractors and the employer
32	Receive notice of discovery of fossils, etc. and instruct regarding disposal
33	Approve cleanliness on the site on completion
35	Receive returns from the contractor
36(1)	Instruct regarding materials and workmanship and tests
37	Require access to works, site, workshops, etc.
38	Receive notice that work and foundations are ready for examination. Direct the uncovering of works
39	Order removal, substitution and re-execution of faulty materials or workmanship
45	Give permission for night or Sunday work
50	Direct searches for defects
51	Vary the works
52(3)	Order work to be executed on a daywork basis
58(1)	Order work on provisional sums
(2)	Order work on prime cost items
(6)	Require submission of quotations, invoices etc.

59(c)   Obtain proof of payment to nominated sub-contractors
60(1)   Receive submission of statements of contract value.

The reasons for the delegation of the engineer's powers to the engineer's representative depend very much on the nature of the project, its location, distance from head office and so on.

In projects constructed overseas many powers which are normally exercised by the engineer in the United Kingdom have to be delegated by necessity to his representative: this is to a large measure dictated by the inherent delays of communication and due to immediate decisions having to be taken on-site for certain tasks. Indeed, the FIDIC and the Overseas Conditions of Contract recognize this fact, and accordingly more powers are delegated to the engineer's representative under these conditions of contract than those delegated under the ICE Conditions. Nonetheless, it is often necessary to extend the powers given under the FIDIC and Overseas Conditions to the engineer's representative by delegating some of the engineer's powers.

When delegating any of the engineer's powers to his representative an internal procedure is evolved on the limits of delegation. For instance, unless the power to vary the works under clause 51 is given, the engineer's representative will not be in a position to make a simple substitution of reinforcement. In order to safeguard the engineer's position from any over-commitment by his representative in this regard certain internal limits are placed on, say, the extent to which the engineer's representative is entitled to vary the works. These limits of delegation that are imposed by an internal procedure, however, may not be disclosed to the contractor.

Duties of the engineer's representative

In civil engineering parlance the engineer's representative is designated normally as the resident engineer. Although the function of the resident engineer under standard conditions of contract is defined as to watch and supervise the works, the duties involved in fulfilling this object are varied and onerous. The resident engineer chosen for the works has to match the scope and nature of the project which he has to supervise, and his seniority dictates to some extent the degree of responsibility that the engineer places on him.

The ICE Conditions of Contract are very clear on one matter – what the resident engineer is not empowered to do: this is to relieve the contractor of any of his duties or obligations imposed on him by the contract. The need for strict adherence to this principle cannot be over-emphasized; it is an edict in management that two parties should not be held responsible for carrying out the same function.

The duties that the resident engineer has to carry out may be expressed in the following terms:

1. The organization of the supervision of the works according to the agreed programme.
2. Agreement of detailed programmes to conform with the overall construction programme.
3. Checking of progress of work at regular intervals in relation to the agreed progress.
4. Supervision of the permanent works and ensuring that the materials and workmanship comply with the specification.
5. Checking of the setting out of the works to ensure that they are in accordance with drawings and the intent of the contract.
6. The examination of method statements for the execution of permanent works and for temporary works proposed by the contractor with special reference to the safety of such proposals.
7. Issue of site instructions to the contractor in the way of clarification of construction drawings and modifications required thereto to the extent of such powers that are delegated to him by the engineer.
8. The admeasurement and agreement with the contractor's site staff of quantities of work executed.
9. The issue of instructions for work to be carried out on daywork, the receipt and checking of statements relating thereto and in prime cost items, before they are incorporated by the contractor into the interim and final payment statements.
10. The agreement of on-account rates for work for which no rates exist in the bill of quantities until such time as they have been fixed
11. Maintenance of records of measurements together with level and dimensions of work as executed and details of any deviations from the working drawings, if any
12. Maintenance of contemporary records in relation to any claims put forward by the contractor and agreement at site on factual data related to claims

13. Channelling of any claims received from the contractor to the engineer. This would include proposed new schedule rates with his recommendation to the engineer
14. Order stoppage of work if it is considered unsafe or is not in compliance with the specification.

The foregoing represents the duties that the resident engineer is called upon to perform normally in relation to the contract.

The resident engineer and his staff are in constant contact with the site agent and his staff during the course of performing these duties. Apart from verbal and written communications exchanged between the respective staffs it is usual to hold a formal progress meeting weekly, where matters such as progress and other matters affecting the construction of the works are discussed. Minutes of such meetings are recorded. Also it is normal for monthly progress meetings to be held at site where representatives from the head offices of both the engineer and the contractor are present; often at the monthly meetings the employer is represented.

Another important factor, intangible though it may be, that contributes to the success of a project is the relationship that the engineer's site staff develops towards their counterparts. It is important for the resident engineer always to be able to act independently, impartially and firmly towards the contractor's site staff, but at the same time it is desirable for him to be understanding and, above all, to have an open mind in regard to any solutions to problems or suggestions that they may have to offer. This is not an easy balance to strike, but is very beneficial for the success of the job when it is done properly.

The resident engineer is responsible for the running of his own organization in addition to his duties in relation to the contract. He has also to report regularly to the engineer on all important matters relating to the contract. Again, such reporting takes the form of weekly and monthly reports in addition to correspondence, telephone conversations and meetings that take place between the head office staff and the resident engineer.

A typical weekly report to be submitted promptly by the resident engineer to the head office should, *inter alia*, include the following:

.Name of client

                 Title of contract
                 Weekly progress report No.
                 for week ending
  Weather: Brief description

Visits:	Visits of head office staff and other persons
Staff movements:	Details of absence on leave or engagement or departure of personnel
Labour:	The total number of men of each class of labour listed separately, shown as an average for the week. Specialist sub-contractor's personnel should be listed separately.
Time lost:	As agreed with the contractor, due to bad weather, strikes and other causes.
Temporary works:	A description of each main section of the work carried out during the week is required.
Permanent works:	A description of each section of the work carried out during the week giving rough estimates of principal quantities, such as volume of concrete placed, volume of excavation and linear metre of drains laid.

Resident engineer

Date

The monthly reports prepared and sent to the engineer by his resident engineer are best submitted in the style of an essay. The report should cover such matters as the weather and the time lost therefrom, men employed, sections of the works, giving a general picture of the project under construction with its progress outlined in relation to the programme of works. This report forms the basis of the monthly report issued to the employer by the engineer.

In the running of the resident engineer's own office several administrative duties have to be undertaken. Above all, the resident engineer must maintain discipline among his staff and endeavour to keep their morale at a high level. If the project is based overseas the resident engineer has to see that his staff, including their families, are properly housed and that there are adequate recreational facilities provided for their well-being. A resident engineer employed in the United Kingdom has to be responsible for certain mundane duties which, if he were in head office, would be done by an administrator as opposed to an engineer. These duties will include a measure of personnel, salaries and accounting administration in respect of staff for which he is responsible.

Resident engineer's organization

The resident engineer's site organization depends on the type of work that is required to be supervised. It follows that the site organization will not be uniform for all sites although the principle governing the organizational structure will always be the same. In a small project the deputy resident engineer will take on the certificate measurement function and the resident engineer will be directly responsible for the technical and contractual aspects of the project, perhaps only assisted by one junior engineer. Likewise for the supervision of hydro-electric schemes, roads, railways, tunnels, dams, power stations and other types of projects, the resident engineer's organization will have to be tailored for its particular needs. The principle that has to be borne in mind when the organizational structure is drawn up is that efficient supervision of the works must be made available for the project to be executed and completed successfully. All too often an employer wishes to save money on site staffing: this is in general not desirable, as proper supervision results in a better end product.

A typical resident engineer's organization chart for a medium-scale civil engineering project is shown in Figure 11.1.

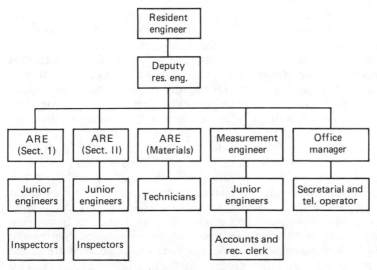

Figure 11.1 A typical resident engineer's organization chart

Chapter 12

Engineer's estimates, methods of pricing and rate fixing

The responsibility for providing an estimate of the capital cost of a project generally falls upon the engineer. The first estimate, which he is often called upon to produce at very short notice, is with minimal data. He may be asked for the cost of establishing a 500-MW thermal power station or for the construction of a 2-mile long tunnel, or for the establishment of a portland cement plant of a specified capacity in any given country. One may term these as 'horseback' estimates which are expected to be of the right order. Such estimates are in a sense difficult to make but, on the other hand, can be, and frequently are, made with surprising accuracy; this is provided the engineer concerned is experienced in the type of project involved. To produce any such estimate he has to have recorded costs of similar projects constructed in the past and a general appreciation of the price levels and escalation indices, ground conditions and technical and economic factors relating to the location at which the particular project is to be established.

The next stage of estimating with which the engineer is concerned is the provision of a capital cost estimate, as an integral part of his feasibility study. During this stage estimates for one or more schemes may have to be provided. These will be produced by the engineer himself or in association with a quantity surveyor. Such an estimate is prepared by taking off approximate quantities and pricing them from previously established rates, indexed as appropriate, or by using price books. Unlike as is often the case on building work, a direct comparison with already established unit rates is not appropriate for civil engineering projects. It is essential therefore for the engineer to evaluate the ground conditions, topography, waterways and other site-related features in arriving at the rates to be used for computing the estimated cost of the project. For instance, on tunnelling work the location of the access shafts or adits and the ground through which the tunnel is to traverse become important functions of the cost. It follows that the programme, including the number of shifts to be worked and the method of construction used, will have a direct and distinct bearing on the capital cost. In the case of such projects and others where costs are closely related to special construction methods it is necessary for the engineer to assume a construction programme and a method of construction to arrive at a realistic cost estimate of a scheme. The cost of the selected scheme together with a contingency element will form the overall cost estimate for the project on which the financial provision or authorization for its implementation will be based.

The third estimate of capital cost prepared by the engineer is that based on the tender documents and drawings prepared for the particular project, and this will refine the estimate prepared for the feasibility study. At this stage the engineer has the details of the work involved set out in the drawings and the specification; where an admeasurement type of contract is adopted the items of work to be carried out are also listed in the bill of quantities. The engineer is in a position at this stage to examine and assess all the unit cost elements of the project and apprise the employer of his estimate of what the true tender value should be.

Methods of pricing and rate fixing

The basic principle underlying the pricing of a tender is that the total cost for the execution of the works, together with a reasonable percentage thereof to cover profit, should be reflected in the tender.

Before attempting to price a tender it is necessary for a contractor to examine all the tender documents received by him and make a thorough study of the requirements set out therein. Thereafter the most economical manner in which the works are to be constructed to fulfil these requirements most economically should be planned.

The pricing of a tender is a complex operation and requires considerable knowledge and experience of estimating before the basic principle involved is fully understood. There are a variety of methods used for pricing a civil engineering tender. However, the fundamental requirements for pricing is that all elements of cost that are to be incurred on the work should be identified and included in the rates and prices that form the tender. These elements of cost may be expressed in the following terms:

1. Labour cost
2. Material cost
3. Plant cost
4. Sub-contract work cost

5. Provisional sums and prime cost items
6. Site on-cost
7. Head office on-cost
8. Profit.

Labour cost

The first task that has to be carried out to determine the labour cost for an operation is to work out an hourly rate applicable to different classes of labour. In the United Kingdom the basic rates payable, together with the conditions applicable for employment of labour in the civil engineering construction industry are contained in the *Working Rule Agreement of the Civil Engineering Construction Conciliation Board for Great Britain*. To these rates other sums as appropriate for specialized working have to be added. In many overseas countries the minimum wages payable to labour are stipulated by the respective government authorities. It is, however, the responsibility of the estimator to evaluate the additional sums payable above the minimum rates so as to attract the right calibre of labour and maintain the required numbers for the duration of the contract. On overseas projects where labour is imported into a country, costs of agency fees payable, air fares, costs of establishing labour camps and other outgoings should be built into the hourly rate for labour.

In building up the hourly rate for labour for work in the United Kingdom some of the elements of cost which should be added to the basic rate of pay would include the following:

1. Non-productive overtime
2. Employer's liability insurance
3. Construction Industry Training Board levy
4. Annual and public holidays
5. Superannuation and redundancy payments
6. Sick pay
7. Guaranteed minimum earnings
8. National Health contributions and life insurances
9. Protective clothing
10. Small tool and boot allowances
11. Subsidy on unearned bonus
12. Time lost due to inclement weather
13. Subsistence allowances where applicable
14. Fares and travelling time
15. Working gangers and charge hands.

The above additional costs could be as much as or more than the basic rate of pay set out in the Working Rule Agreement.

Material cost

The cost of materials to be used for pricing is generally based on quotations obtained for its supply. To this cost the estimator should add the cost of transportation to site, unloading, handling and distribution within the site. An allowance should also be made for waste.

Plant cost

Contracting organizations as a general rule maintain plant hourly costs for the operation of standard items of plant. In arriving at such a rate the following factors are considered:

1. Depreciation costs
2. Maintenance and repair costs
3. Cost of insurance
4. Taxes.

To this hourly cost, depending on the type of plant and the particular project, the following items of cost are added:

1. Freight or transport costs to and from site
2. Cost of tyres for transport equipment
3. Fuel and consumables
4. Plant operators' costs.

Sub-contractors' cost

On civil engineering projects a fair measure of work is carried out by both domestic sub-contractors chosen directly by the main contractor and those nominated by the engineer. The main contractor normally provides site facilities such as the use of canteens, offices, workshops, medical centre, sanitation, water supplies and electricity, and also is responsible for the general supervision of the sub-contractors' work. These costs have to be determined and a percentage is normally added to the sub-contractors' costs to cover these items in the event they cannot be absorbed under site on-costs.

Provisional sums and prime cost items

In respect of some provisional sums and all prime cost items entered in the bill of quantities, the tenderer is called upon to quote a percentage for charges and profit against those items. In quoting the percentages the contractor should assess the costs likely to be incurred by him in respect of such items and allow for them in the quoted percentages.

Site on-cost

The site on-costs are the general costs that are incurred by the contractor for his site establishment, temporary works, services, general site installations and other items which are not directly built into the cost of the items of permanent works. The items of work which fall into this category may be classified as follows:

1. Contractor's site offices, including maintenance, heating, cleaning, office furniture, equipment and telephones
2. Site supervision and staff, including foremen and walking gangers
3. Temporary works
4. Welfare facilities
5. Insurances
6. Survey equipment and chainmen
7. Electricity and water supplies
8. Access road, hardstandings and fencing
9. Clearing site and reinstatement
10. Contractor's yard, workshops and stores
11. Watching and lighting
12. Control of traffic
13. Safety and site security
14. Personnel transport
15. Other site obligations imposed by the conditions of contract or the specification and not separately identified in the bill of quantities.

The costs for the foregoing are computed and distributed as a percentage addition on the direct cost of measured items in the bill of quantities, unless they have been separately identified in the bill.

Head office on-cost and profit

Head office on-costs are the contractor's head office costs that are incurred for the supervision and direction of the contract as well as those he is obliged to incur over and above the site on-costs. These include directors' fees, head office salaries, office rents, furniture, equipment and establishment charges, financing charges and the like. A contractor would express the turnover of the particular contract as a percentage of his gross turnover and apply this percentage as an addition to the other costs of the project. In addition, the costs of obtaining a performance bond, contract insurances and any special insurance he may have to take out for overseas projects will be included in this percentage unless such costs are directly reimbursable in terms of the contract.

The contractor, having worked out his percentage for head office on-costs, will then add his percentage for profit to arrive at the final rates and prices to form his tender.

Rate fixing

The whole question of rate fixing was a matter of relative simplicity until after the Second World War. In the United Kingdom, for example, the large programme of thermal generating stations in the context of constant advancement of power station plant gave effect to many new, as compared with contract-rated, items.

A new approach was then required to arrive at new rates analogous to those in the contract. A paper was published in 1963 to cover what was then a new facet in the area of rate fixing. Much of the explanatory notes which are still valid have been taken from that paper.*

The basis for valuation of ordered variations and fixing rates for such work in the United Kingdom flow from clauses 52(1) and 52(2) of the ICE Conditions of Contract, where these Conditions are incorporated into a contract. The principles outlined in these clauses are also generally valid where other standard forms of contract are used.

Clause 52(1) requires the engineer to value varied work of similar character carried out under conditions similar to those contained in the bill of quantities at rates and prices contained therein where such are applicable. Where the work is not of a similar character or not carried out under similar conditions then it is to be valued on the basis of the rates and prices contained in the bill of quantities in so far as these may be reasonable, failing which a fair valuation is to be made. The value of the varied work is to be ascertained by the engineer in consultation with the contractor, and should there be a disagreement in this regard the responsibility falls on the engineer to determine and notify the contractor the appropriate rate or price in accordance with the principles laid down in this clause.

* 'Rate Fixing in Civil Engineering Contracts' by C. K. Haswell, ICE Paper No. 6662, February 1963, and Discussion, January 1964.

Clause 52(2) empowers the engineer to fix rates or prices which are reasonable and proper for varied work when, in his or the contractor's opinion, the rates or prices contained in the tender are rendered unreasonable or inapplicable due to such a variation.

The conditions under which a new rate is called for may vary quite considerably. In the simplest case, for instance, the engineer may decide that a leaner or richer concrete mix than specified shall be used in a section of the works; as will be seen later, it is relatively straightforward to arrive at the correct new or analogous rate. On the other hand, unforeseen physical conditions, deprival of access or the construction of extra works, such as a totally new and different structure from any in the tender documents, involves some complexity in arriving at new rates. These difficulties may be reduced to a minimum if a rational and consistent policy is adopted by the engineer in the procedure and method of fixing new rates.

Basis of rate fixing

In preparing a tender a contractor has first to work out the total of his on-cost. The next step is to price each item of the bill of quantities, and this is done by evaluating the labour, material and plant contents; to this is then applied the on-cost, usually in a pro-rata manner.

We may express this as a formula:

$$R = L\left(1 + \frac{l}{100}\right) + M\left(1 + \frac{m}{100}\right) + P\left(1 + \frac{p}{100}\right) + X$$

where R = schedule rate
L = labour content
l = percentage on-cost on labour
M = material content
m = percentage on-cost on material
P = plant content
p = percentage on-cost on plant
X = any miscellaneous costs taken separately.

There are, of course, many variations that can be applied when using the above formula. For instance, a labour percentage only may be used, keeping the materials and plant percentages nil; or the labour, material and plant may be taken together and an overall percentage applied for spreading the on-cost.

A satisfactory interpretation of his responsibility is for the engineer to determine, for new or substituted items, the rate which the contractor would have inserted against that item had the necessary data been available at the time of tender. To implement this it is obviously necessary for the engineer to be familiar with the usual practice adopted by contractors in preparing their tenders.

It will be appreciated that it is not difficult to price any work which involves a simple operation independent of site conditions and without relation to the remainder of the work. An example of this would be the fixing of a door and frame, or the additional cost of formwork for a hole or rebate in a concrete structure. It is also relatively easy to price an entire building contract where there is little interdependence between items. On the other hand, when dealing with major civil engineering projects the contractor of necessity plans the construction as an entity where interavailability and continuous employment of plant is a vital factor. This may be considered in some ways analogous to an airline, which depends for its very existence upon keeping aircraft in the air for the maximum possible periods, carrying the maximum permissible loads.

The importance of plant is illustrated by the fact that a civil engineering tender must be built up from a carefully planned plant programme. Around this plant the necessary labour is phased to keep it working and to supply both the temporary and permanent materials needed for construction purposes in the correct quantities and at the correct time. All this calls for properly qualified executive, engineering and administrative staff, for with construction, as in many other matters, time is a function of money.

The four elements of cost have been illustrated as materials, labour, plant and on-cost, and some notes on each of these are now given as affecting the process of rate fixing.

Direct material cost

The material cost is ascertained in tendering by obtaining quotations from one or more suppliers. Similarly, for a new or substituted item either the cost of the materials used can be ascertained where a schedule of basic prices of materials is included in the contract or quotations have been obtained, or in some cases the actual materials will be in stock.

This is, of course, the easiest of the components of a rate to evaluate, as the only imponderables are the percentage of waste involved and the number of uses obtained from temporary materials.

Direct labour cost

This item is evaluated by the contractor's gauging, from his own experience, the gang size required and the time taken to carry out any operation. This is in effect an exercise in output per man-hour. Published estimating handbooks are, of course, of assistance in this, for no one has experience of every trade. However, these references are used cautiously and not as a rigid rule, for by their very nature they cannot take into account all the circumstances of a particular situation which may, for example, call for two extra men in a gang, or which may lower the output due to the difficulties inherent in access to the site of the work. When jobs overseas are priced, adjustments as appropriate should be made to reflect a realistic output from labour envisaged to be employed. The engineer, in assessing the labour content of a new or substituted item, will follow similar lines.

Direct plant cost

The question of plant, as has been mentioned, is of fundamental importance and requires much thought and planning during the tender stage. This item of cost is arrived at in a manner similar to the labour item, envisaging the available and suitable plant and choosing from it that which is best to use in terms of output, as well as considering the economics of bringing on special plant for any substantial operation. The engineer in assessing the plant content for a new or substituted item will follow lines similar to those adopted in the tender, taking into account variations in output of plant and any new plant that may be required especially to carry out the work involved in such items.

On-cost (site and head office)

This last item is probably the most difficult to assess. Construction of general temporary works (including water and electricity services, fencing, watching, etc.); hutting; camps and offices; wet time; general labour (black gang, etc.); foremen; engineers; administrative staff; chainmen; general stores, and so on, have all to be included if not otherwise allowed for in the bills of quantities.

All these and, of course, head office costs and profit margins, are the indirect costs and overhead charges comprised in the total on-cost. It will be appreciated that while assessment of the direct cost in each case is a relatively straightforward operation, this is not the case as far as on-cost is concerned. This requires much expert knowledge of contracting and is subject to the requirements of each firm's policy at the time of tender. The on-cost is estimated for the whole of a tender and then spread pro rata over the three items of direct cost in any manner suitable to the tenderer; this is effected by percentage additions, as shown in the formula set out in this chapter. In fixing rates for new or substituted items the percentages used in the tender are applied, unless a particular element of cost in the make-up of the on-cost is not applicable for that particular item. In the latter case an adjustment will have to be made with due regard to the make-up of the on-cost used in the tender.

Procedure for rate fixing

At the outset of a contract the engineer will require to know the percentage on-cost for labour, materials and plant in the unit rates and prices in the contract. The best method is for the engineer to ask the contractor for a breakdown in, say, three substantial and differing items. It must be stated that the contractor is entitled to refuse to provide the data requested; in such cases the engineer will have to make his own breakdown of unit contract rates. In general, the contractor usually provides the required information since it is to his advantage to do so.

There are various ways in which a new rate is initiated. Often it is left to the contractor to apply to the engineer when it appears to the former that a new rated item is justified. This in itself is not at all unreasonable, always provided that the engineer issues a revised bill of quantities where there occur major changes in the character of any section of the works. Further, he should advise the contractor of any substantial item such as occurs where a change in the material content is specified after the contract is placed. It is probable that all such matters are best dealt with through the resident engineer on-site.

In so far as new or substituted items are concerned it is for the contractor, through his agent, to propose any new rate to the resident engineer, together with the 'make-up'. It is for the engineer, through the resident engineer, to accept or reject the principle of a new rate. If it is accepted he fixes a suitable figure which is used on an 'on-account' basis in the interim certificate measurement.

In this connection it is prudent for the resident engineer to be conservative in fixing the 'on-account' rate. This obviates concern on the part of those contractors who consider such rates in the interim certificate measurement as the minimum.

This procedure presupposes that the rate is not finally settled before the work is put in hand. Obviously it is better if the rate can be settled beforehand and firm endeavour to this end should be made. However, it is often not practicable to achieve this object.

The next step in the procedure is that the resident engineer considers the proposal and makes his recommendation to his head office. He should forward the contractor's make-up, as well as his own, together with suitable comment relating to actual site conditions and so on.

From time to time rate-fixing meetings are held, usually at the engineer's head office. Representatives of the contractor's head office are present, and the agent and the resident engineer should also attend. At these meetings all new rates are fixed by the engineer with the contractor's acceptance, if not always his agreement.

It is axiomatic in general – and in particular under the ICE Conditions – that the engineer shall fix the new rates. However, in practice it is rare for new items not to be agreed to by the contractor in discussion at the rate-fixing meeting.

It is a prerequisite to the successful conclusion of these meetings that the engineer shall be represented by competent and experienced engineers versed particularly in the practice of construction of civil engineering works. This is a very important point and as such should be emphasized.

Boundary of rate fixing

It is not always easy for the engineer to decide where the boundary of rate fixing lies. It is wise if there is general agreement between the employer and the engineer on this point before any work is put in hand: the employer should appreciate and acknowledge the scope of the engineer's power in this respect. This is, in effect, a proper appreciation of the conditions of contract.

It should be remembered that if any item for rate fixing is rejected by the engineer it can always be presented by the contractor in the form of a claim proper after the end of the contract period or earlier.

In considering any case for a varied rate made by the contractor, the engineer should reject it if he considers that it is properly held to be included within the tender rate. Equally it should be rejected if it lies within the wider issue of what is termed a contract risk.

In many cases the matter has to be treated with caution; to illustrate this an example may be of assistance. A large number of bearing piles require to be driven; difficulty in driving the piles is experienced by the contractor initially and he invokes clause 12 of the ICE Conditions of Contract covering unforeseen physical conditions. In the event, the last, say, 9000 out of a total of 10 000 piles may well drive more easily than could reasonably have been forecast during the tender period. If an extra to the tendered rate under ICE clause 12 had been accepted at the time, the contractor would have scored unfairly at the cost of the employer.

It is not unusual, therefore, for the engineer to be wary in the acceptance of new rates. It is often best for contractual matters in relation to progress of the contract as a whole to be dealt with at the completion of the works.

Chapter 13

Claims

In the context of a civil engineering contract normally a claim means a demand by a contractor for payment of an item or items of work carried out by him on behalf of the employer for which a readily identifiable amount cannot be ascertained under the terms of the contract. Such a claim is always made upon the employer but under standard forms of contract it is first considered by the engineer and, should his decision be disputed, it is adjudicated by arbitration or in the courts of law.

Claims fall into two main categories:

1. *Contractual claims*
   These claims are made under the provisions of the contract entered into between the employer and the contractor. It follows that the contractor is entitled to receive payment under the contract for valid claims made under this category.

2. *Extra-contractual claims*
   These claims are those which are either not made under the conditions of contract or are those allegedly made under them but considered legally unenforceable by the employer on the basis that they do not fall within their provisions.

   Consequently payment in respect of such claims is considered to be outside the framework of the contract. Claims which fall into this category often arise because a contractor has suffered a loss which cannot be attributed to a fault of either party within the terms of the contract.

   A decision to recompense a contractor in respect of extra-contractual claims may be taken by an employer on what he considers to be a moral obligation or it may be offered merely in order to avoid an admission of liability. Claims of this nature are often examined in this light and, where possible, *ex-gratia* payments are granted to settle such claims.

   Claims by a contractor can arise due to an altered condition of contract in the widest sense, an initial inadequacy of the contract documents by the way of lack of clarity and omissions, a lack of communication between the engineer and the contractor or only for the reason that the contractor wishes to recover sums of money to increase profits or reduce losses: such losses may be due to underpricing or to his own shortcomings in constructing the works.

In a civil engineering contract claims are invariably submitted by the contractor in the first instance for consideration by the engineer. The engineer has a duty under the contract to resolve disputes and consider claims in an independent and impartial manner without showing bias towards either the employer or the contractor. Under standard forms of contract the contractor must abide by the engineer's decision on any dispute that may arise in connection with the works unless and until the dispute is referred to arbitration and an award is made and published on the reference by the arbitrator.

In the context of a contractor's claim, whether it be for time or money or both, the engineer has to make a decision both on the validity of the claim and on its quantification, and this becomes a question of whether any particular claim may be placed in the category of a contract risk or in the category of a contract responsibility.

Claims procedure

Under the ICE Conditions of Contract a contractual claim may be submitted to the engineer at any time during the currency of the contract and until three months after the date of the maintenance certificate issued by the engineer. The claim must be for something to which the contractor considers himself entitled under the terms of the contract up to the date of the maintenance certificate. Contractual claims, being those which arise from specific clauses of the conditions of contract, should refer to the clause number(s) under which the claim is made so as to obtain initial acceptance for it. The majority of claims on civil engineering contracts, incorporating the ICE Conditions of Contract, are, however, made under the following clauses:

Clause 12. Adverse physical conditions and artificial obstructions

Clause 44. Extension of time

Delay claims arising out of clauses 7, 12, 13, 14, 27, 31, 40, 42 and 59B.

Clause 51. Ordered variations

Clause 52. Valuation of ordered variations and rate fixing.

Claims under clause 12

Perhaps the most extensively used clause for submission of claims in the civil engineering industry is the clause 12 of the ICE Conditions of Contract or a clause which has a similar effect in other conditions of contract adopted for civil engineering works. Essentially, this clause is a safety-valve that begins to operate when an experienced contractor encounters conditions during the execution of the works that he could not have reasonably foreseen at the time of submitting his tender. The existence of this clause is invaluable in that a contractor is not asked to price for a risk that he cannot foresee; the employer likewise is not called upon to accept a higher tender price in respect of something which may never happen. In some overseas countries a clause sensibly similar to clause 12 of the ICE Conditions of Contract is purposely omitted. This is done in the belief that the existence of such a clause *per se* invites a claim; but if the intent of the clause is followed by both the contractor and the engineer the equitable nature of the clause can be said to be of benefit to both parties to the contract.

However, it must be said that the clause, although well intended, is far too often resorted to by the less reputable contractors, which has tarnished the clause and its principles. Claims under clause 12 of the ICE Conditions of Contract arise from unforeseeable adverse physical conditions or artificial obstructions which are encountered on the works. Those that are considered unforeseeable at the time of tender by an experienced contractor may be expressed in the following terms.

Adverse physical conditions

The following physical conditions may warrant a claim for additional payment provided, of course, they were unknown at the time of tender:

Wells
Waterways
Geological faults
Ground conditions different from those anticipated
Subsidence
Unpredictable water tables.

Clause 12 does not rank weather conditions or conditions due to the weather as a physical condition giving rise to a claim. It follows that exceptional rain, storms or floods, or landslips arising therefrom, do not entitle a contractor to additional reimbursement under this clause.

An unforeseen physical condition may be stated to have been encountered if the ground conditions are materially harder than those anticipated, on the one hand, and if they are substantially softer than anticipated, on the other. This situation may and at times does occur on tunnel contracts. As an example, on a tunnel driven for a hydro-electric project the borehole information provided on the line of the tunnel indicated that the tunnel was to be driven in rock. In this particular case there was no other evidence available to the contractor to deem otherwise. Accordingly, the contractor arranged the tunnel to be driven by machine and attendant rock-tunnelling equipment. During the driving operation soft soil was encountered for some 180 metres of the tunnel, the total length of which was some 3000 metres. This caused the contractor a delay of some ten months on a contract period of 36 months. In this particular case the contractor's claim for additional reimbursement and for extension of time under a contract clause similar to clause 12 of the ICE Conditions of Contract was successful. In a like manner, a clause 12 claim situation may occur if the contractor had been led to believe that a soft ground-tunnelling situation existed and the method of excavation had to be changed to rock tunnelling for part of the tunnel drive.

Artificial obstructions

The following are some examples of artificial obstructions that are man-made which can give rise to a claim provided they were not apparent with proper search or brought to the notice of the contractor at the time of tender:

Sewers and manholes
Services
Adits, shafts and tunnels
Old structures
Concrete, brickwork, masonry and other man-made obstructions.

The magnitude of the obstruction governs whether any of these items warrant a claim to be made under this clause. A small obstruction could be paid for under another item in the bill of quantities provided for a class of work similar to the obstruction met. As an example, the contract rate for rock excavation may be an appropriate rate to be applied when a part of old mass concrete foundations has to be demolished and carted away. Likewise, the removal or relocation of existing services could be paid for on the basis of daywork. Again, it is possible to deal with obstructions by issuing a variation order and to value such works without recourse to a clause 12 claim.

Under clause 11 of the ICE Conditions of Contract the contractor is deemed to have examined and inspected the site and carried out any other relevant investigations and satisfied himself of the conditions thereon before submitting a tender. It follows that for a clause 12 claim to be justified it is for the contractor to demonstrate that the physical conditions or artificial obstructions met during the execution of the works could not reasonably have been foreseen by him as a contractor experienced in the type of work for which he had submitted a tender.

Equally important is for the contractor to demonstrate that the condition in fact cost more to overcome than could properly have been foreseen.

Claims under clause 44

Clause 44 of the ICE Conditions of Contract is the clause under which the contractor can seek and the engineer can grant extensions of time for completion of the works. At the outset the point must be made that an extension granted under this clause does not automatically entitle the contractor to additional costs incurred by him, if any, for the period of the extension.

Under sub-clause (1) the duty is placed upon the contractor to deliver to the engineer full and detailed particulars of any claim for extension of time to which he considers himself entitled. A claim can arise due to a variation ordered under clause 51(1), increased quantities referred to in clause 51(3), for any other cause of delay referred to in the conditions of contract, exceptional adverse weather conditions or other special circumstances of any kind whatsoever. The contractor is required to deliver the particulars relating to a claim within 28 days after the cause of the delay has arisen, or as soon thereafter as is reasonable in all the circumstances, in order that the claim can be investigated at the time.

The engineer's duties under this clause are set out in sub-clauses (2), (3) and (4).

Under sub-clause (2) the engineer is required to grant an interim extension of such time as he considers is due on the information then available to him, that is, on the basis of the particulars submitted by the contractor, or in the absence of such particulars (or a claim), based on all the circumstances known to him.

Under sub-clause (3) the engineer has a duty to assess any extensions of time to which he considers the contractor is entitled at the due date for completion, whether or not the contractor has made a claim for extension of time. This situation, of course, will not arise if the works or the relevant section of the works are substantially complete by the due date for completion.

Under sub-clause (4) the duty falls upon the engineer to review all circumstances pertaining to an entitlement to an extension of time and finally determine and certify the overall extension of time (if any) that he considers due upon the issue of the completion certificate for the whole or sections of the works, as the case may be. In determining this extension the engineer is not empowered to decrease any extension that had been previously granted.

The engineer is bound to notify the contractor of any extensions that are granted. He is also required specifically to notify the contractor of any claims for extensions of time to which he decides the contractor is not entitled. Sub-clause (3) of the clause requires that both the employer and the contractor be notified should the engineer consider that the contractor is not entitled to an extension of time at the due date for completion; such notification will enable the employer to start deducting liquidated damages if he so elects. In practice, the engineer normally informs the employer of extensions of time granted or refused at any stage of the contract because the time for completion is a matter of major importance on any contract.

Delay claims

As stated above, the granting of an extension of time does not in itself qualify for payment in respect of delays. However, if it can be demonstrated by the contractor that the delays do not fall within the category of a contract risk (*q.v.*) and that they are 'real' delays for which payment is specifically allowable or implied by other clauses in the contract, then the contractor would be entitled to payment in respect of extra costs incurred by him.

A 'real' delay may be stated as a period during which the contractor is unable to deploy his men and plant, in a manner that the intended output or progress of work can be achieved in relation to the critical path of the agreed programme. Some of the clauses in the conditions of contract which specifically refer to payment for delays are:

Clause 7(3):	Delay and extra cost due to delay in issue of drawings and instructions.
Clause 12(3):	Delay and extra cost arising out of adverse physical conditions and artificial obstructions.
Clause 13(3):	Delay and extra cost arising out of the engineer's instructions, including those given by him to clear ambiguities or discrepancies on the several documents forming the contract referred to in clause 5.
Clause 14(6):	Delay and extra cost in relation to the engineer's consent on methods of construction, design criteria and allied matters covered by clause 14.
Clause 27(6):	Delays and extra cost attributable to variations in relation to the Public Utilities Street Works Act.
Clause 31(2):	Delays and extra cost arising out of affording facilities for other contractors.
Clause 40(1):	Delay and extra cost arising out of suspension of work.
Clause 42(1):	Delay and extra cost arising out of the failure to give possession of site at the agreed dates.
Clause 59B(4):	Delay and extra cost arising out of a forfeiture of a sub-contract.

In addition to the above, the contractor has the right to claim for extra costs for delays and disruption on the basis of any of the other clauses in the conditions of contract. The notices of claims for additional payment are always lodged under clause 52(4) of the conditions of contract.

Claims under clause 51

Clause 51 deals with ordered variations, and, from the point of view of claim entitlement, this is an all important clause. Points to note are that under this clause the engineer's powers to vary parts of the works are limited to those which in his opinion are necessary or desirable for the completion or functioning of the works. When variations of a large nature are made it may be prudent for the contractor to ensure that the engineer is acting within his authority. Also it is a requirement of the clause that ordered variations should be in writing or, if a verbal order is given, that such be also confirmed in writing. There is no requirement under this clause for a special procedure to be followed by the engineer when variations are ordered. A letter instructing the contractor to do some work or the issue of a working drawing amending a previous one will be as valid as a formal numbered variation order issued by the engineer. Provided proper authorization of the variations has been obtained, the contractor is entitled to be paid for them under the contract.

Claims under clause 52

This clause deals with valuation of variations at sub-clause (1) and rate fixing by the engineer at sub-clause (2).

For the purposes of valuation of ordered variations the contractor generally puts forward proposals for their costs to be ascertained by the engineer. The principles to be applied for such valuations are set out in sub-clause (1) and the powers given to the engineer to fix rates and prices for varied work is set out at sub-clause (2). The manner in which rate fixing is dealt with on civil engineering contracts is contained in the previous chapter. A claim proper arises under this clause only when the contractor is dissatisfied with the rates or prices notified to him by the engineer, and consequently claims higher rates or prices.

Notification of claims

Whatever form of contract is adopted, a claim, when it occurs, has to be notified to the engineer at the appropriate time for such a claim to have a chance of success. In the ICE Conditions of Contract the manner in which the notification of a claim is to be lodged and the procedure relating thereto is dealt with in six paragraphs bearing letters (a) to (f) of sub-clause 52(4). From a point of contractual claims this clause ranks high in importance to a contractor. The essential features of this clause are set out below.

Paragraph (a) deals with the notification of claims for higher rates and prices than those notified to him by the engineer pursuant to clauses 52(1), 52(2) and 56(2). Clauses 52(1) and 52(2) were discussed previously in this chapter: clause 56(2) relates to any increase or decrease to the contract rates and prices that have been notified to the contractor by the engineer consequent to the contract rates or prices being considered by him to be unreasonable or inapplicable, because of significant changes in actual quantities from those in the bill of quantities. A point to note is that a notice of claim under this paragraph must be given within 28 days of receiving an engineer's notification pursuant to clauses 52(1), 52(2) or 56(2).

Paragraph (b) deals with the notification of claims for additional payment pursuant to any clause in the conditions of contract other than in respect of sub-clauses (1) and (2) of Clause 52 which are covered in paragraph (a) above. Notification of a claim under this paragraph is to be made as soon as reasonably possible after the happening of the events giving rise to the claim. The contractor is also required to maintain such reasonable contemporary records as are necessary to support any subsequent claim.

Paragraph (c) allows the engineer, without admitting liability, to instruct the contractor to keep such reasonable contemporary records and gives the engineer the right to inspect them.

Paragraph (d) lays down the procedure to be adopted for submitting accounts and particulars after the notice of a claim under clause 52(4) has been given.

Paragraph (e) points out that should the contractor fail to comply with any of the provisions of this clause he shall be entitled to payment only to the extent the engineer has not been prevented from or substantially prejudiced in investigating the claim by such failure.

Under paragraph (f) of this sub-clause the contractor is entitled to be paid the amounts considered due on claims by the engineer through interim payment certificates. The engineer has to be provided with sufficient particulars to determine the amount that is due and if they are insufficient to substantiate the whole claim the contractor would be entitled to only the part of the claim which is substantiated to the satisfaction of the engineer.

As an example, the procedure that is to be adopted for the notification of a claim for encountering adverse physical conditions and artificial obstructions under the terms of clause 12 of the ICE Conditions of Contract is given below.

Contractor's notification

Clause 12(1) of the conditions of contract deals, *inter alia*, with the method of notifying a claim under this clause. It requires that:

1. The notice of an intention to claim be given to the engineer under clause 52(4) of the conditions of contract.

2. The contractor specifies in the notice the physical condition or artificial obstructions encountered.
3. With the notice, if practicable, or as soon as possible thereafter, details of the anticipated effects, the measures that are being taken or proposed to be taken and the extent of the anticipated delay in or interference with the execution of the works are required to be provided.

Engineer's action

Clause 12(2) deals with the action required by the engineer in response to the contractor's notification: the engineer may, if he thinks fit:

1. Require the contractor to provide an estimate of the cost of the measures he is taking or proposes to take.
2. Approve in writing such measures with or without modification.
3. Give written instructions as to how the physical conditions or artificial obstructions are to be dealt with.
4. Order a suspension under clause 40 or a variation under clause 51.

Contractor's entitlements

Should the engineer decide that the claim is acceptable in whole or in part, then under clause 12(3) he is required to:

1. Determine any extension of time for completion under clause 44.
2. Certify payment of reasonable costs for additional work done.
3. Certify payment of reasonable costs for additional constructional plant used.
4. Certify payment of reasonable costs incurred by the contractor by reason of any unavoidable delay or disruption of working suffered by him.
5. Certify payment of a reasonable percentage addition in respect of profit on costs of additional work done and plant used, but not on delay and disruption costs.

Conditions foreseeable

Under clause 12(4) the engineer is required to inform the contractor in writing as soon as he has decided that conditions under which the claim is made could have been reasonably foreseen by an experienced contractor. It should, however, be noted that under this clause the contractor is entitled to be paid for work carried out by him in respect of any variations ordered by the engineer in response to the initial clause 12(1) notice served by the contractor upon the engineer; the value of such work is to be ascertained under the provisions of clause 52.

Preparation of claims

A fundamental prerequisite for the preparation of a contractual claim is that an entitlement for it should be evident from one or more clauses incorporated into the contract. Having identified the relevant clauses it is then necessary to collate all the facts and evidence available to the contractor in support of his claim so as to demonstrate its validity to the engineer. The first step that the contractor has to take when a claim situation arises is to notify the engineer of his intention to claim under the relevant clause or clauses of the conditions of contract within the time limits stipulated therein. Brief particulars of the claim should be set out in the notification unless the full and detailed particulars relating to the claim can be sent with it. The comprehensive particulars should be submitted to the engineer as soon as this is practicable and with due regard to the particular requirements of the clauses under which the specific claim is made.

Once the principle of the claim has been accepted by the engineer it is necessary to present a fully prepared quantified claim. This would still be necessary even if the engineer did not accept the claim in principle, and it is the intention of the contractor to seek a formal decision of the engineer under clause 66 of the ICE Conditions of Contract or a similar clause in another form of contract. Should the contractor be dissatisfied with such a decision then the dispute can be referred to arbitration, in which event the fully prepared quantified claim will form a part of the points of claim prepared for the arbitration.

A formal claim has to be set out clearly and be well prepared so that the basis of the claim and the evidence in support of it are presented in a manner that is readily understood by the engineer, or, if the need arises, by the employer and lawyers. As supporting evidence, a case history of the events leading to the cause of the

6

claim should be given. The claim should also tabulate and set out the relevant documents, events and technical considerations that the contractor relies upon to prove his entitlement for the claim.

Some of the documents from which facts and information for the preparation of the claim may be elicited are listed below:

1. Full set of contract documents including the priced bill of quantities and any qualifications accepted by the employer as forming an integral part of the contract.
2. Contract drawings.
3. Working drawings including all revisions and bar-bending schedules.
4. Contract correspondence.
5. Variation orders, engineer's instructions and site instructions.
6. Minutes of meetings at head office and site levels.
7. Measurement files.
8. Interim certificate measurements.
9. Engineer's certificates.
10. Site-progress photographs.
11. Correspondence on nominated sub-contractors and suppliers.
12. Site records.
13. Programme and progress charts.
14. Site survey and level notebooks.
15. Method statements and contractor's proposals.
16. Plant and labour records.
17. Site investigation reports and soils information made available at the time of tender.
18. Further site investigations or tests carried out during construction or, if necessary, after the completion of the works.
19. Cost and finance records.
20. As-completed record drawings and sketches.

The manner in which a claim is presented depends on the type of claim, its magnitude and also on the organization and the person or persons responsible for its preparation. A typical claim submission may follow the format set out below:

Title page
Table of contents
Part 1: Contract particulars
Part 2: Claim particulars
Part 3: Evaluation of claim
Part 4: Summary.

Where appropriate, documents that are relied upon in support of the claim would be attached as appendices.

Chapter 14

The nature of arbitration and related procedure

Arbitration may be defined as a procedure for the settlement of a dispute by referring to the judgement of a selected person or persons outside the established courts of law.

In its broad sense, arbitration is used as an instrument for effecting settlements in many aspects of modern life. These may range from conciliation in industrial disputes on a national basis to international arbitration in such matters as, for instance, fishing rights. However, it is here proposed to deal with arbitration as it affects the practising civil engineer during the course of his work, mainly in relation to contracts carried out under the ICE Conditions of Contract.

It may be of interest to consider the reasons why this form of dealing with disputes has arisen. Arbitration procedures existed in the ancient city-states of Greece and in medieval England; perhaps also the Judgment of Solomon may be cited as an early form of arbitration award. It is only in fairly recent times, however, that this procedure has been developed in the British Isles. The development of English law was largely concerned with the principle that no tribunals should be permitted which were not reponsible to the judiciary. Presumably, however, the increasing complexity of disputes in a highly industrialized country, turning as they frequently do on matters of a highly specialized and technical nature, gave rise to the need for an independent arbitrator. Also there was the wish of the parties not to air their secrets in open court. Such an arbitrator would generally be someone with expert knowledge of the matters in dispute and would, thus, more easily be able to decide on complicated technical issues. Nevertheless, any prospective arbitrator who has listened to a High Court or County Court judge delivering judgment in an action involving technical evidence will realize that the law is not all that much an ass, to quote Dickens.

In the United Kingdom the law as to arbitration in England and Wales is embodied in the Arbitration Act 1950 as amended by the Arbitration Acts 1975 and 1979. The 1950 Act was a consolidation of the previous Acts of 1889 to 1934. The main Act governing arbitration is Scotland is the Arbitration (Scotland) Act 1894 and that applicable to Northern Ireland is the Arbitration Act (Northern Ireland) 1937.

In English law, the jurisdiction of the courts cannot be entirely ousted and there are various grounds, which will be examined later, under which an arbitration award may be brought under the control and correction of the court.

For a dispute to be resolved by arbitration there must be an agreement between the parties for such a reference. This assumes that in the contract in question there is incorporated an arbitration clause; otherwise the resolution of a dispute between the parties is normally by litigation in the law courts of the particular country in question.

It may be stated that even where an arbitration clause exists either party is at liberty to start proceedings in court. However, the other party can apply to the court for a stay of such proceedings on the basis of the existence of an arbitration clause: in the United Kingdom the court will generally grant such a stay if the dispute falls within the terms of the arbitration clause. One of the grounds on which a stay may be refused is when another party who is not a party named in the arbitration clause is also involved in the dispute, but this is now affected by the Arbitration Act 1975 (see *Lonrho* v. *Snell*, 1978). A reason for the almost universal inclusion of an arbitration clause is that the parties are seeking a method of resolving disputes which is final, swift and private. Although an arbitration certainly can be said to be private the will has to exist in the minds of the parties to the dispute as well as the arbitrator if a decision is to be swift and final. This is because the Respondents may be in a position to delay proceedings by frequent appeals back to the judicial system, although in UK and international contract practice this does not happen often. Further, under the Arbitration Act 1979 the involvement of the courts in the process of arbitration has been considerably restricted from those flowing from the Act enacted in 1950.

From the experience of having had technical cases decided in the High Court it may be said that in most disputes greater confidence can be placed in an arbitration than a court case on a subject matter that is entirely of a technical and techno-financial nature, provided that the arbitrator appointed is technically qualified and experienced. Such an arbitrator should also understand legal procedures and should be aware of any failure or inability of either party to comply with proper requirements as to evidence and procedure.

The question also arises as to the number of arbitrators that should be stipulated to hear a dispute. Single arbitrators are almost invariably used in civil engineering contracts in the United Kingdom. The Institution of Civil Engineers' Conditions of Contract provides for arbitration by a single arbitrator. Under the Arbitration Act 1950 it is deemed that the reference shall be to a single arbitrator where no other mode of reference is provided.

The Overseas Conditions of Contract also provides for a single arbitrator, while the International Conditions of Contract (FIDIC) leaves the option open for either a single arbitrator or for more than one arbitrator to be appointed in accordance with the rules of the International Chamber of Commerce. The rules of the ICC provide for disputes to be settled by a sole arbitrator or by three. When a dispute is referred to three arbitrators each party is required to nominate one arbitrator for confirmation by the ICC Court and the third arbitrator, who will act as the chairman of the arbitration tribunal, shall be appointed by the court unless the parties have provided that the arbitrators nominated by themselves may agree on a third within a fixed time limit. Failure to agree on the third arbitrator by the arbitrators within the time limit empowers the Court to appoint the third arbitrator. Where arbitrators are appointed by the parties, they should not be advocates for the party who appoints them, but at all times should act in a fair and unbiased manner.

Although in the authors' experience a single arbitrator is considered to be more satisfactory for the speedy resolution of a dispute, certain advantages are cited in support of the adoption of three arbitrators. One such advantage is that it enables sound engineers who may have had extensive experience of arbitration to act under the guidance of the third, who might be a senior arbitrator experienced in the matters in question or an experienced lawyer. This can have the advantage of reducing the need for the arbitration tribunal to obtain further advice on matters of a legal or technical nature.

In civil engineering contracts an arbitration clause is normally included. The inclusion of such a clause is also normal practice in mechanical and electrical engineering, building and other related contracts.

If we look first at the causes of a dispute it will be seen that we should, at the outset, review the various documents which form the basis of the contract. Then we shall be better equipped to deal with the subject.

Unfortunately, it is true to state that a very usual cause of a dispute is an initial inadequacy of the contract documents by way of lack of clarity, definition and omission, and indeed errors, leaving the onus of interpreting something inadequate in the contract upon the engineer. If the documents were perfect, fewer disputes would arise. It is also true to say that all too often, despite the care taken in the preparation of documents and drawings, a dispute may be taken to arbitration solely for the reason that there has been a breakdown in communication between the engineer and the contractor and either or both have taken up an uncompromising position which, in all probability, could have been resolved by argument and discussion.

Often one must have sympathy with the engineer. An employer, albeit with good reason, may delay taking the vital decision of authorization for a project. Once taken, the employer may find it expedient to debar the engineer from sufficient time properly to carry out the pre-tender tasks of feasibility study, economic design and planning, site investigation, topographical survey and finally the preparation of the contract documents and drawings.

Quite a common cause of dispute is the taking of an erroneous decision by the engineer in interpreting the contract. Such mistakes are often very expensive for the employer.

Another reason for which it may also be considered to be unfortunate for a dispute to arise is that it may be initiated by a contractor only for the reason of wishing to gain sums of money without due reason. The causes are many and varied: to increase profits; to reduce losses due to underpricing; bad management and/or lack of drive; back luck or bad forecasting in the rate of striking and so on. Any contractor's case in this context rests often on some error or anomaly in the conditions of contract, specification, bill of quantities, drawings or instructions.

Again, the claimant (and this is usually the contractor) may give notice of arbitration with the object only of inducing an employer to settle a dispute in his favour, that claimant having no intention of ultimately proceeding to arbitration because he knows or has been advised by his lawyer that there is no adequate case to present.

The dispute in civil engineering contracts

General

A contract is an amalgam, as it were, of duties and rights of the parties to that contract. Unlike a contract for sale of goods, a civil engineering contract is almost certain to encounter circumstances somewhat different from those expected when the tender documents are prepared. This may be due to a variety of reasons; ground conditions, weather, economic circumstances, shipping delays, currency problems – to name but a few. The result is that some modification of the contract will be necessary during the course of construction. This is fertile ground for disputes, but goodwill and common sense on the part of the engineer and the contractor should obviate serious dispute, except in extreme circumstances.

In this context the fundamental questions of cause and effect of a dispute should now be examined. The cause may be any one of those mentioned earlier: the effect is the difference of opinion between that contained in the engineer's decision and that of the employer or, more usually, the contractor in the interpretation of the contract. In the majority of cases the dispute stems from differences of opinion in the element of duty rather than a right between the engineer and the contractor in relation to contract risks and contract responsibilities: these have been discussed in Chapter 4. It is essential to appreciate the fundamental differences between these two elements in dealing with disputes. In this context we may define these two elements in the following terms. A contract risk is something which should have been allowed for in a tender and for which the contractor is financially at risk. A contract responsibility is something which the contractor has to carry out within the terms of the contract, but for which he takes little or no financial risk.

Perhaps the most difficult decision before an engineer is one in relation to clause 12 of the ICE and the FIDIC Conditions of Contract for dealing with unforeseen physical conditions. If the matter in question is classed as a contract risk no additional payment is due to the contractor. If, on the other hand, the matter is classed as a contract responsibility, the extra sums thereby occasioned are due to the contractor.

Dealing with disputes

It is a principle of a contract into which the ICE Conditions of Contract are incorporated that all differences of opinion or disputes arising out of the contract shall be settled by the engineer. The employer and the contractor are required to give immediate effect to the engineer's decision, which remains final and binding unless and until the decision is referred to arbitration and the arbitrator has published an award. This principle is equally valid under the FIDIC Conditions of Contract for civil engineering construction, with the added requirement for parties dissatisfied with the engineer's decision to make an attempt first to reach an amicable settlement before the commencement of arbitration.

Although the requirement to make an attempt to reach an amicable settlement prior to arbitration is not expressly provided for in the ICE Conditions of Contract this is also a course of possible action that can, in certain cases, settle a civil engineering dispute without recourse to arbitration. Either the contractor or the employer, or occasionally the engineer, can call in an independent engineer: the latter should be a civil engineer of some eminence, well practised in the particular field of work involved and in contract matters. It is best if that person has also acted as an arbitrator. Wisdom and knowledge brought by a fresh and unfettered mind can often bring a dispute to the correct solution and agreement between the parties concerned.

The dispute

Let us now examine the processes whereby a dispute in a civil engineering contract may find itself referred to arbitration.

A common means of establishing a civil engineering contract is for an employer to seek competitive tenders from a number of civil engineering contractors.

The employer represents the body for whom the project is to be constructed and, although he may undertake the role of the engineer, he frequently appoints a consulting engineer to act in the capacity of 'the engineer'.

The contractor is the person carrying out the work of the construction and the two parties to the contract are the employer and the contractor. It has been usual in the United Kingdom since 1955 for civil engineering contracts – whether admeasurement, lump sum or cost/reimbursement – to be placed with contract documents and drawings using the ICE Conditions of Contract.

The arbitration clauses incorporated into the Conditions of Contract for Civil Engineering Construction of the Institution of Civil Engineers, together with those in the Overseas and the International forms are given in the Appendices to this chapter. All stipulate in the respective clauses that the decision of the engineer on a dispute shall, in respect of any matter referred to him, be final and binding on the parties to the contract until such matter is referred to and finally settled by arbitration. This stipulation is an important device whereby a contract could proceed without interruption, although either party to the contract may be dissatisfied with the engineer's decision. Sometimes this important stipulation is not made in conditions of contract drawn up by overseas countries. It is the authors' view that for civil engineering contracts this would be considered a serious omission if continuity of construction during arbitration is to be maintained.

The arbitration process

The advantage of arbitration over an action in the courts especially manifests itself when the issues are largely those of a technical nature where experience is involved.

Disputes may be dealt with more quickly by arbitration and the hearing may be arranged to suit the parties.

The hearing itself is private and thus not open to the public, which may be important in matters of a confidential, technical or financial nature.

It is often thought that arbitration proceedings are less expensive than a court action. However, the recent tendency to employ leading counsel, added to the costs resulting from lengthy hearings as well as other expenses, makes this rather doubtful in some cases.

Conduct of the arbitrator

Before discussing the procedure for the hearing it may be of interest to draw attention to several important aspects of the conduct of a prospective arbitrator. The arbitrator must be scrupulous in observing certain rules of behaviour which he will ignore at his peril.

First, in the words of the famous seventeenth-century Chief Justice Coke, 'no man may be a judge in his own cause'. It follows that an arbitrator may be disqualified if he has any undeclared financial interest, however small – for example, even a few shares – in one of the parties to the dispute. If this interest is known and accepted by both parties, however, there will probably be no difficulty.

Second, on the maxim of Lord Hewart that 'Justice should not only be done, but should manifestly and undoubtedly be seen to be done', any contact with one of the parties to the exclusion of the other, however, innocent, may nullify the proceedings. In one case an arbitrator visited the site of the dispute in the company of one party only and, although it was not established that the case was discussed, the award was set aside by the courts. It is important, therefore, to emphasize that an arbitrator must act at all times with the utmost impartiality and put himself, like Caesar's wife, above suspicion.

Reference to arbitration

A dispute is referred to arbitration in terms of the arbitration clause contained in the contract. The ICE Conditions of Contract stipulate in its arbitration clause that the person to be appointed as the arbitrator may be agreed between the parties, failing which he is to be nominated on the application of either party by the President of the Institution of Civil Engineers. The conduct of the reference under this clause is to be in accordance with the ICE arbitration procedure. Such reference is deemed to be within the meaning of the Arbitration Act 1950. For works situated in Scotland the contract has to be interpreted in accordance with Scots law, and the ICE's arbitration procedure for Scotland has to be followed.

On international contracts the manner in which a dispute is to be referred to arbitration is set out in the particular forms used for such contracts. The International form (FIDIC) requires arbitrations to be conducted under the rules of the International Chamber of Commerce Court of Arbitration unless otherwise specified in the contract. An alternative set of arbitration rules frequently adopted for the settlement of international disputes are those produced by the London Court of International Arbitration or the United Nations Commission on International Trade Law (UNCITRAL). These rules provide for the parties to an arbitration to designate an administration authority which normally will be an arbitration body such as the International Chamber of Commerce or others established in the countries concerned. The appointment of an arbitrator under the ICE Overseas Conditions is treated in a manner similar to the United Kingdom ICE form, except that the appointment authority of the arbitrator, failing agreement by the parties, is to be stated in the form of tender. Any reference to arbitration on International or ICE Overseas forms are deemed to be submissions to arbitration within the meaning of the arbitration laws of the country specified in the contract.

Preliminary meeting

Once an arbitrator has been appointed it is usual for him to hold a preliminary meeting before the actual hearing, except in simple cases where this is unnecessary. When a reference to arbitration is made under clause 66 of the ICE Conditions of Contract the holding of a preliminary meeting is mandatory under Rule 10 of the ICE Arbitration Procedure (1983), while under its corresponding procedure for Scotland this is optional. In the former case, too, should the arbitrator agree, this requirement can be waived.

The proceedings at this preliminary meeting are somewhat in the nature of the proceedings on a Summons for Directions in an action in the High Court, and the purpose of a preliminary meeting is to identify such matters as may be cleared before the hearing, and to issue the necessary directions for conducting the arbitration.

These matters may include the fixing of the date and place of hearing, the requirement or otherwise for questioning of witnesses under oath, the ordering of a transcript of the proceedings and whether the parties will be represented by counsel. Neither party is obliged to be represented by a solicitor or counsel, but this will depend on the nature of the dispute. The other matters at the preliminary meeting to be resolved are:

1. The confirmation of the procedure to be adopted for the arbitration.
2. Order for Directions.
3. Particulars of any party's claim or counterclaim, as the case may be.
4. Discuss the discovery and inspection of documents.
5. The inspection of property and things:
   (i) By parties
   (ii) By the arbitrator.
6. Fix dates for the delivery of pleadings, i.e. points of claim and points of defence.
7. Exchange of proof of evidence.
8. Any other matters required for or to expedite the hearing.

Perhaps it might be useful to explain some of the above matters, although most of them are self-explanatory.

Order for Directions
This is the arbitrator's order to both parties for proceeding with the arbitration. It covers, *inter alia*, such matters as pleadings and dates for delivery, discovery of documents, and the date and place of the hearing.

Pleadings
The pleadings are formal documents, usually settled by counsel or a solicitor, in which each party sets out his case and answers the other party's case so as to define the areas of dispute. A party is not normally permitted to raise any issue at the hearing which has not been pleaded.

The pleadings are in effect the points of claim from the claimant; the points of defence in which the respondent replies to the points of claim; and the reply which answers the defence if a further answer is necessary. Where the defence also includes a counterclaim, the reply will include or take the form of a defence to counterclaim.

Discovery of documents
Discovery means the disclosure of all documents which are or have been in the possession or control of each party and which are in any way relevant to the issues in the arbitration.

A party may refuse to disclose a document on the grounds that it is privileged. The most important privileged documents are communications between a party and his solicitors for the purpose of obtaining legal advice and also communications between the solicitor and third parties in connection with a contemplated dispute. Legal opinions and expert technical reports, for the purposes of the proceedings, are usually privileged. A party may waive the privilege if it is wished to put such documents in evidence.

In fixing the date and place of the hearing the arbitrator has sole discretion, subject to anything in the agreement. He must, however, act reasonably in this respect and must not act precipitately to catch one of the parties unaware or the award may be set aside (cf. *Oswald* v. *Grey*, 1855).

Any refusal by either party to attend the hearing after reasonable notice may empower the arbitrator to proceed with the hearing without that party, i.e. *ex parte*.

The hearing
The arbitrator has power to decide the conduct of the case and the procedure is for the hearing to follow the rules of evidence as in a civil court.

The general procedure, is, therefore, for the claimant to open his case (with or without an advocate) and then to call his own witnesses, who are thereafter cross-examined by the respondent or his counsel and then re-examined. The same procedure is then adopted by the respondent, who then sums up his case. During the hearing the arbitrator may also question witnesses. The claimant, who has the right to the last say in the case, then replies to the respondent's summing-up and this concludes the hearing.

In the examination by a party of his own witnesses – known as the evidence or examination-in-chief – no leading questions may be asked, i.e. questions which lead the witness to the reply which the questioner seeks. For example, 'Did you notice the poor quality of the concrete in the column?' is a leading question, whereas, 'When you examined the column, was there anything you noticed?' is not. Leading questions to the opposing party's witnesses are permitted in cross-examination.

There are certain exceptions to the rule against leading questions in examination-in-chief such as names, occupations and so on for identification of things or for matters not in dispute. In re-examination the questions must be confined to those raised in cross-examination, although the arbitrator may allow fresh issues to be raised by either party.

Regarding evidence, the arbitrator should never receive evidence from one party without the knowledge of the other. If he received a communication from one party he should immediately inform the other or the award may be set aside. The question of evidence itself is a highly complex matter and the arbitrator is advised to be circumspect.

If the arbitrator relies on evidence which is immaterial or rejects evidence which is material, he may be equally blameworthy. This is a highly difficult position for an arbitrator and requires considerable experience and judgement. It is important to see that no material evidence is excluded, and it is prudent for an arbitrator, who is not a lawyer, to be patient and listen to both parties, although he should use his judgement to see that the proceedings are not unnecessarily protracted and hence costly.

The award

Formal requisites of award

The award itself must be made to comply in point of form with any directions contained in the submission.

If these directions are not complied with the award will be considered invalid unless those directions are of a totally immaterial nature.

Although a parol award may be considered as valid, unless the submission requires otherwise, the award should be made in writing.

The arbitrator may not delegate the making of his award to another person. However, there is no objection to an arbitrator employing a legal adviser to assist in drafting the award and to give the necessary advice in this direction. The adviser should, of course, be independent in the same way as is the arbitrator. The authors' view is that the wise arbitrator will always obtain legal advice to draw up the award: this includes the matter of costs.

It is usual practice in the making of an award for the arbitrator to sign it and for that signature to be witnessed. However, it is not unusual for an award to be made in duplicate, one of which is signed and thus becomes the original award and this is delivered to the party taking up the award: another copy is available for the other party if required. At all events, no stamp duty is required on the award. Another way of dealing with the matter which is preferable is that there is only one award, and this is taken up by the party on payment of the arbitrator's fees.

There is some confusion as to the question whether the arbitrator shall make and publish his award. If this provision is made in the submission, the award is not valid unless it is made and published. However, this does not seem to be of great importance, since it may be held that the award is considered to be published as soon as the arbitrator has in fact made a complete award. Once made and published, the award must be published to the parties. The arbitrator does this by notifying the parties that his award has been made and published and that it is ready to be delivered.

It should be noted that it is sometimes stipulated by a submission that the award should be ready to be delivered by a specified date: this means what it ways. The fact that it is not actually delivered by that date will not affect its validity. On the other hand, if the stipulation is that the award shall be delivered by a particular date then actual delivery is essential to its validity.

It is usual for the arbitrator to retain the award until his fees have been paid by the party taking it up. This course has been approved by the court and each arbitrator is advised to follow it: otherwise he may never be paid.

Substantive requisites of award
In order that the award shall be valid it must be final, certain, consistent and possible, and it must decide upon all the matters submitted and no more than the matters submitted.

We have said that the award must be final and where this is not so it will be referred back to the arbitrator or set aside. It follows that the arbitrator must be careful to ensure that the wording is a final decision on all matters requiring his determination.

The arbitrator may award one sum generally in respect of all monetary claims submitted, unless in the submission there is a clear intention of the parties to award separately, or there is some legal necessity for doing so as, for instance, to determine the right to costs. Where several matters are in issue the award may have to decide upon each of them.

Time for making award

Unless the arbitration agreement expresses a contrary intention, it is possible for the award to be made at any time. If there is undue delay in making an award, the remedy is the removal of the arbitrator.

Directions in the award

Implied powers to give directions

Items in general which come under this category are rarely relevant in civil engineering contracts since they relate to such matters as that an arbitrator is not able to bind a man's liberty, although he has the same power as a court to order specific performance of any contract. Again, an arbitrator may not direct an act to be done contrary to law.

Directing payment of money

An arbitrator may in general fix the time for payment to be made and indeed he may direct that such a payment be made by instalments. An arbitrator may even direct that payment is made upon a Sunday.

Interest

As an axiom we may say that an arbitrator may award interest by virtue of his implied authority to follow the ordinary rules of law.

The question of interest as affecting the sum or sums of money awarded by an arbitrator may conveniently be divided into two elements: the first of these relates to any interest which the arbitrator may include in his award. In this connection it is not unusual for the advocate representing the plaintiff to plead for interest, and the arbitrator will, at the time of making his award, decide whether to accept or modify the plea and at least take note of the defence advocate's counter-argument.

Where the question of interest has not been mentioned during the hearing the arbitrator will have to decide from when and at what rate interest should be paid, if any. As a general rule, it would be fair to state that interest might be granted at, say, the average base rate, and that it should be awarded from the time approximating that at which claims are normally settled in a civil engineering contract.

Minimum Lending Rate was abandoned by the Bank of England in August 1981, and while this institution continues to exert a measure of influence over interest rates through their control of the money supply, the British clearing banks now fix their rates in relation to their base rates. The borrower of money is usually asked to pay 1 per cent to 5 per cent above the base rate, depending on his borrowing status. The current base rate may be established by reference to one of the major clearing banks.

With regard to interest subsequent to the award, any sum or sums directed to be paid in the award carry interest in the same way and at the same rate as a judgment debt unless the award directs otherwise.

Costs

The arbitrator has full discretion as to the costs of the reference since the arbitration clause expresses no contrary intention.

The term 'costs of the reference' includes all the expenses properly incurred by the parties before and during the course of the hearing before the arbitrator.

The arbitrator has full power to deal with the costs, and indeed this he must do in his award, except in the case where the parties to a reference are agreed as to who shall pay the costs and have so informed the arbitrator.

Although we may say that in general terms the costs follow the event, experience in arbitration has shown that the legal view in the awarding of costs is somewhat complicated from the point of view of the civil engineering arbitrator. The latter should, therefore, take the advice of his legal adviser, as has been mentioned previously.

The complication in awarding of costs may be illustrated by the claimant who, say, is claiming £100 000 and is awarded only £1000. One way of looking at it is that the claimant should pay 99 per cent of the costs. On the other hand, if he had not entered into arbitration proceedings, he would have received nothing, so that one may argue that he should pay only, say, 75 per cent.

Effect of an award

The award is in effect a final judgment on all the matters referred to the arbitrator.

The arbitrator may not reopen or recall the final award once it is made, except to correct a clerical mistake. He can only alter it if it is remitted to him by the court.

The court has no power to alter or amend an award; it can only set it aside or remit it to the arbitrator.

Enforcement of awards

Scope and effect of the section

An award may be enforced if made pursuant to a written arbitration agreement by leave of the High Court or a judge thereof in the same manner as a Judgment or Order. This is on the assumption that the award is in a form which is capable of being enforced in the same manner as a Judgment.

It should be noted that the validity of an award is not necessarily affected by the fact that a party after entering into a submission becomes bankrupt before the execution of the award.

It is also worth noting that the award will not be enforced as a judgment under section 26 of the Arbitration Act 1950 and the successful party will be left to bring an action on the award in the following areas: where the submission and agreement to refer are by parol; where the award is declaratory only, that is, where it ascertains only the amount to be paid and not any liability in law to pay; and also where the validity of the award, or the right to proceed upon it, is not reasonably clear.

Enforcement of award by action

On the question of enforcement it is necessary for the plaintiff to prove:
1. The making of the submission, or else the making of a contract which contains an arbitration clause, followed by the arising of a dispute within that clause.
2. Appointment of an arbitrator or arbitrators in accordance with the arbitration agreement, unless the arbitral tribunal is named by the agreement itself.
3. The making of the award.
4. The amount awarded has not been paid, or otherwise the award has not been performed.

Defences to action on award

It should be noted that the defendant cannot in any action on an award plead as defence the misconduct or irregularity of the arbitrator. His proper course, if these grounds exist, is to have the award set aside.

An action to enforce an award must be brought within six years of the date on which cause of action accrued unless the submission is by an instrument under seal.

Again, the defence in an action may plead that the arbitrator has exceeded his jurisdiction.

It is not a good defence, if an arbitration agreement does not provide for a time in which the award should be made, to plead that it was not made within a reasonable time.

Challenging the award

An award made pursuant to a proper arbitration agreement is final between the parties unless the agreement expresses a contrary intention. However, in general there is no appeal and an award will be binding unless some degree of invalidity can be demonstrated.

Although at common law there is no power to remit, this power is contained in the Arbitration Act, and if any matter is remitted to the arbitrator it must be considered and determined by him. It follows that this reconsideration brings with it the question of redetermination.

The Act allows for the power of the court to set aside an award on the question of misconduct by an arbitrator or an umpire, as the case may be. In this context the term misconduct should be taken in its widest sense, including mistakes in law or fact admitted by the arbitrator. It is worth noting that this misconduct justifying intervention by the court may take place at any stage between appointment and entering upon a reference, during a reference, or in the making of an award.

The following may be taken as an illustration:

An arbitrator was asked by the parties at the conclusion of the hearing (a) to issue his award in relation to a specified number of items only in draft form, so that the parties might consider whether or not to ask for a case stated, and (b) to make no award as to costs until the parties had had time to consider the remainder of his award and had made submissions to him at a further hearing. By inadvertence, the arbitrator issued his award in final form as to a number of the specified items, and purported to make an award as to costs. *Held* by Wynn J. that the arbitrator was guilty of misconduct and that the award should be set aside and remitted to the arbitrator so that he might comply with the parties' request (*Marples Ridgeway & Partners Ltd* v. *CEGB,* 1964).

It was Litvinoff, the famous erstwhile foreign minister of Russia, to whom was attributed the well-known phrase 'Peace is indivisible'. This may be said to be analogous to civil engineering and indeed to other disciplines.

Civil engineering covers research, development, feasibility, environment, survey, design and so on; lastly, it covers construction and maintenance in which the contract system is of significance.

As an apodosis to the contract of construction is the resolution of dispute: in the limit, as it were, this matter of dispute is settled by arbitration.

An endeavour has been made to illustrate the salient features of the final settlement of a contract by arbitration.

For the optimum the engineer must be of integrity; he must be technically wholly competent; he must be experienced in the disciplines to be adopted; he must be thoroughly versed in all aspects of matters of contract as has been outlined; and he must be legally adequate. Overall, he must be fair to both parties in interpreting the contract: the decision he makes must be impeccable.

The arbitrator must possess all the attributes of the engineer in full measure with greater depth of experience and, of course, knowledge of arbitration. He has to review and pass judgment on the engineer's decision. Ideally, the engineer should be well versed in arbitration, since he should know how a decision would fare in arbitration.

The ICE Arbitration Procedure

The first standard procedure published for the conduct of an arbitration flowing from ICE Conditions of Contract was the ICE Arbitration Procedure (1973). This document, which was eventually published in March 1976, contained the Procedure itself together with notes for guidance, the forms needed for a reference to arbitration and appendices. The procedure covered the steps leading up to the hearing, including the appointment of the arbitrator, preliminary meetings and draft orders for directions. The notes for guidance which were intended to give a background on arbitration to the parties in dispute included sections on the powers of the arbitrator, conduct of the hearing, rules of evidence and the preparation of the award. Clause 66(1) of the fifth edition of the ICE Conditions of Contract published in 1973 made the use of the Procedure optional, becoming binding upon the parties only if they agreed to its adoption or if the President of the ICE so directed. In the latter event the arbitrator also had to agree and the parties should not dissent for the procedure to be used. Until the first standard procedure was published the arbitrator, on appointment, had to glean from various textbooks the correct procedure, and this was not wholly satisfactory.

The current arbitration Procedure was published in 1983 and replaces the earlier 1973 document. The current document is also designed for use with clause 66 of the fifth edition of the ICE Conditions of Contract. In terms of the latest version of clause 66 of the ICE form the use of the ICE Arbitration Procedure is mandatory (unless, of course, the parties agree otherwise in writing) instead of the optional provision that existed in the fifth edition of the ICE Conditions of Contract before it was revised in 1985.

The ICE Arbitration Procedure (1983) and the revised version of clause 66 of the ICE Conditions of Contract issued by the standing joint committee of the ICE Conditions of Contract in March 1985 as appearing in the 1986 reprint are set out in full in Appendices 14.1 and 14.2, respectively. Appendix 14.2 also sets out the arbitration clauses contained in the FIDIC Conditions of Contract for Civil Engineering Construction (fourth edition 1987) and those from the ICE Conditions of Contract for Overseas Work Mainly of Civil Engineering Construction (first edition, 1956).

ICE Arbitration Procedure (1983)

The ICE Arbitration Procedure (1983) comprises nine parts which contain 27 rules. This document first sets out the procedural matters that are to be undertaken in a reference to arbitration and then deals with the powers available to and the procedures that may be followed by an arbitrator.

Matters relating to the reference and the appointment of an arbitrator are in Part A, which comprises four rules. This part sets out largely the process for obtaining the appointment of an arbitrator. Rule 4 details the manner in which new and connected disputes may be added to the arbitration.

The relevant powers of the arbitrator are shown in Part B, which contains Rules 5 to 9. The arbitrator's powers to control the proceedings, to order protective measures, to order concurrent hearings, at the hearing and to appoint assessors or to seek advice are the topics covered by these rules.

The procedure to be adopted before the hearing is covered under Rules 10 to 14 in Part C. Rule 10 makes it mandatory for the arbitrator to summon a preliminary meeting, although provisions exist within this rule whereby the parties are not bound to attend under certain conditions. Rule 11 deals with pleadings and discovery; Rule 12 covers the calling of procedural meetings prior to the hearing; Rule 13 deals with orders which may have to be given by the arbitrator to streamline the procedure at the hearing while Rule 14 covers the manner in which summary awards are to be made.

The procedure to be adopted at the hearing is explained at Part D, which includes Rules 15 and 16. Rule 15 deals with the manner and order in which the case is to be presented and the issues to be decided upon by the arbitrator. Rule 16 empowers the arbitrator to regulate the way in which evidence is placed before him.

Part E covers the matters the arbitrator has to do after the hearing, which essentially concerns the publishing of his award. Rules 17, 18 and 19, which deal, respectively, with the making and publishing of the award; giving reasons for the award; and the requirement to notify the arbitrator forthwith if a party applies to the High Court for leave to appeal against the award or for other purposes constitute the three rules included under this part.

Parts F, G and H are three special procedures available to the parties and the arbitrator where traditional procedures are considered to be inappropriate. The first of these is Part F – short procedure, which contains Rules 20 and 21. This procedure is intended to provide an opportunity for the parties to obtain a quick decision on the matter in dispute at minimal cost. Although this can apply to any dispute it is considered more suited where the amount of money at issue is small.

Rule 20, applicable to Part F, deals with the manner in which the short procedure can be adopted and Rule 21 covers other matters related to the use of this procedure.

Part G, which sets out the special procedure for experts, contains Rules 22 and 23. This procedure is intended to provide the means of obtaining a decision quickly on matters which depend on evidence of opinion given by experts. Under this procedure the experts retained by each party, without the intervention of lawyers, review the evidence at a meeting called by the arbitrator. Following the meeting the arbitrator makes a binding award on issues of fact. Rule 22 details the working of the special procedure for experts and Rule 23 covers the order on costs the arbitrator may make relating to the reference conducted under this procedure.

Part H, which is the third special procedure, deals with interim arbitration, that is, arbitration which takes place before the completion of the works. Rule 24 is the only rule in this Part. Prior to the revision of clause 66 of the fifth edition of the ICE Conditions of Contract in March 1985, a reference to arbitration under this clause had to await the completion of the works unless they were in respect of disputes arising under clause 12 (adverse physical conditions or artificial obstructions) or in withholding by the engineer of any certificate.

Rule 24 had been designed for interim arbitrations falling within the old 1973, clause 66(2), of the ICE Conditions of Contract. The new version of clause 66 gives the right to a reference to arbitration on any dispute arising out of an engineer's decision under this clause before the completion of works. Provided an arbitration consequent to such a reference commences before the substantial completion of the works, the parties are free to choose between the interim procedure under Rule 24 or the provisions of the main procedure, as appropriate to the issues involved, for the conduct of the arbitration.

Part J is devoted to miscellaneous provisions of the Arbitration Procedure and contains Rules 25, 26 and 27. The definitions that apply to the Rules are given in Rule 25. These are declatory and are included for the avoidance of doubt; Rule 26 sets out the conditions under which the ICE Procedure shall apply to the conduct of the arbitration. Rule 27, which is the last rule of the ICE Arbitration Procedure, entitled 'Exclusion of Liability', excludes the liability of the ICE, its servants, agents and the President for any act, omission or misconduct in connection with any appointment made or any arbitration conducted under the Procedure.

The Institution of Civil Engineers' Arbitration Procedure (1983) for England and Wales

Part A. Reference and Appointment

Rule 1. Notice to Refer

1.1   A dispute or difference shall be deemed to arise when a claim or assertion made by one party is rejected by the other party and that rejection is not accepted. Subject only to Clause 66(1) of *the ICE Conditions of Contract* (if applicable) either party may then invoke arbitration by serving a *Notice to Refer* on the other party.

1.2   The Notice to Refer shall list the matters which the issuing party wishes to be referred to arbitration. Where Clause 66 of the ICE Conditions of Contract applies the Notice to Refer shall also state the date when the matters listed therein were referred to the Engineer for his decision under Clause 66(1) and the date on which the Engineer gave his decision thereon or that he has failed to do so.

Rule 2. Appointment of sole Arbitrator by agreement

2.1   After serving the Notice to Refer either party may serve upon the other a *Notice to Concur* in the appointment of an Arbitrator listing therein the names and addresses of any persons he proposes as Arbitrator.

2.2   Within 14 days thereafter the other party shall

(a)   agree in writing to the appointment of one of the persons listed in the Notice to Concur or

(b)   propose a list of alternative persons

2.3   Once agreement has been reached the issuing party shall write to the person so selected inviting him to accept the appointment enclosing a copy of the Notice to Refer and documentary evidence of the other party's agreement.

2.4   If the person so selected accepts the appointment he shall notify the issuing party in writing and send a copy to the other party. The date of posting or service as the case may be of this notification shall be deemed to be the date on which the Arbitrator's appointment is completed.

Rule 3. Appointment of sole Arbitrator by the President
3.1   If within one calendar month from service of the Notice to Concur the parties fail to appoint an Arbitrator in accordance with Rule 2 either party may then apply to the President to appoint an Arbitrator. The parties may also agree to apply to the President without a Notice to Concur.

3.2   Such application shall be in writing and shall include copies of the Notice to Refer, the Notice to Concur (if any) and any other relevant documents. The application shall be accompanied by the appropriate fee.

3.3   The Institution will send a copy of the application to the other party stating that the President intends to make the appointment on a specified date. Having first contacted an appropriate person and obtained his agreement the President will make the appointment on the specified date or such later date as may be appropriate which shall then be deemed to be the date on which the Arbitrator's appointment is completed. The Institution will notify both parties and the Arbitrator in writing as soon as possible thereafter.

Rule 4. Notice of further disputes or differences
4.1   At any time before the Arbitrator's appointment is completed either party may put forward further disputes or differences to be referred to him. This shall be done by serving upon the other party an additional Notice to Refer in accordance with Rule 1.

4.2   Once his appointment is completed the Arbitrator shall have jurisdiction over any issue connected with and necessary to the determination of any dispute or difference already referred to him whether or not the connected issue has first been referred to the Engineer for his decision under Clause 66(1) of the ICE Conditions of Contract.

Part B. Powers of the Arbitrator

Rule 5. Power to control the proceedings
5.1   The Arbitrator may exercise any or all of the powers set out or necessarily to be implied in this Procedure on such terms as he thinks fit. These terms may include orders as to costs, time for compliance and the consequences of non-compliance.

5.2   Powers under this Procedure shall be in addition to any other powers available to the Arbitrator.

Rule 6. Power to order protective measures
6.1   The arbitrator shall have power

    (a) to give directions for the detention storage sale or disposal of the whole or any part of the subject matter of the dispute at the expense of one or both of the parties

    (b) to give directions for the preservation of any document or thing which is or may become evidence in the arbitration

    (c) to order the deposit of money or other security to secure the whole or any part of the amount(s) in dispute

    (d) to make an order for security for costs in favour of one or more of the parties and

    (e) to order his own costs to be secured

6.2   Money ordered to be paid under this Rule shall be paid without delay into a separate bank account in the name of a stakeholder to be appointed by and subject to the directions of the Arbitrator.

Rule 7. Power to order concurrent Hearings
7.1   Where disputes or differences have arisen under two or more contracts each concerned wholly or mainly with the same subject matter and the resulting arbitrations have been referred to the same Arbitrator he may with the agreement of all the parties concerned or upon the application of one of the parties being a party to all the contracts involved order that the whole or any part of the matters at issue shall be heard together upon such terms or conditions as the Arbitrator thinks fit.

7.2   Where an order for concurrent Hearings has been made under Rule 7.1 the Arbitrator shall nevertheless make and publish separate Awards unless the parties otherwise agree but the Arbitrator may if he thinks fit prepare one combined set of Reasons to cover all the Awards.

Rule 8. Powers at the Hearing
8.1   The Arbitrator may hear the parties their representatives and/or witnesses at any time or place and may adjourn the arbitration for any period on the application of any party or as he thinks fit.

8.2   Any party may be represented by any person including in the case of a company or other legal entity a director officer employee or beneficiary of such company or entity. In particular, a person shall not be prevented from representing a party because he is or may be also a witness in the proceedings. Nothing shall prevent a party from being represented by different persons at different times.

8.3   Nothing in these Rules or in any other rule custom or practice shall prevent the Arbitrator from starting to hear the arbitration once his appointment is completed or at any time thereafter.

8.4   Any meeting with or summons before the Arbitrator at which both parties are represented shall if the Arbitrator so directs be treated as part of the hearing of the arbitration.

Rule 9. Power to appoint assessors or to seek outside advice

9.1   The Arbitrator may appoint a legal technical or other assessor to assist him in the conduct of the arbitration. The Arbitrator shall direct when such assessor is to attend hearings of the arbitration.

9.2   The Arbitrator may seek legal technical or other advice on any matter arising out of or in connection with the proceedings.

9.3   Further and/or alternatively the Arbitrator may rely upon his own knowledge and expertise to such extent as he thinks fit.

Part C. Procedure before the Hearing

Rule 10. The preliminary meeting

10.1   As soon as possible after accepting the appointment the Arbitrator shall summon the parties to a preliminary meeting for the purpose of giving such directions about the procedure to be adopted in the arbitration as he considers necessary.

10.2   At the preliminary meeting the parties and the Arbitrator shall consider whether and to what extent

(a) Part F (Short Procedure) or Part G (Special Procedure for Experts) of these Rules shall apply

(b) the arbitration may proceed on documents only

(c) progress may be facilitated and costs saved by determining some of the issues in advance of the main Hearing

(d) the parties should enter into an exclusion agreement (if they have not already done so) in accordance with S.3 of the Arbitration Act 1979 (where the Act applies to the arbitration)

and in general shall consider such other steps as may minimise delay and expedite the determination of the real issues between the parties.

10.3   If the parties so wish they may themselves agree directions and submit them to the Arbitrator for his approval. In so doing the parties shall state whether or not they wish Part F or Part G of these Rules to apply. The Arbitrator may then approve the directions as submitted or (having first consulted the parties) may vary them or substitute his own as he thinks fit.

Rule 11. Pleadings and discovery

11.1   The Arbitrator may order the parties to deliver pleadings or statements of their cases in any form he thinks appropriate. The Arbitrator may order any party to answer the other party's case and to give reasons for any disagreement.

11.2   The Arbitrator may order any party to deliver in advance of formal discovery copies of any documents in his possession custody or power which relate either generally or specifically to matters raised in any pleading statement or answer.

11.3   Any pleading statement or answer shall contain sufficient detail for the other party to know the case he has to answer. If sufficient detail is not provided the Arbitrator may of his own motion or at the request of the other party order further and better particulars to be delivered.

11.4   if a party fails to comply with any order made under this Rule the Arbitrator shall have power to debar that party from relying on the matters in respect of which he is in default and the Arbitrator may proceed with the arbitration and make his Award accordingly. Provided that the Arbitrator shall first give notice to the party in default that he intends to proceed under this Rule.

Rule 12. Procedural meetings

12.1   The Arbitrator may at any time call such procedural meetings as he deems necessary to identify or clarify the issues to be decided and the procedures to be adopted. For this purpose the Arbitrator may request particular persons to attend on behalf of the parties.

12.2   Either party may at any time apply to the Arbitrator for leave to appear before him on any interlocutory matter. The Arbitrator may call a procedural meeting for the purpose or deal with the application in correspondence or otherwise as he thinks fit.

12.3   At any procedural meeting or otherwise the Arbitrator may give such directions as he thinks fit for the proper conduct of the arbitration. Whether or not formal pleadings have been ordered under Rule 11 such directions may include an order that either or both parties shall prepare in writing and shall serve upon the other party and the Arbitrator any or all of the following

- (a) a summary of that party's case
- (b) a summary of that party's evidence
- (c) a statement or summary of the issues between the parties
- (d) a list and/or a summary of the documents relied upon
- (e) a statement or summary of any other matters likely to assist the resolution of the disputes or differences between the parties

Rule 13. Preparation for the Hearing

13.1   In addition to his powers under Rules 11 and 12 the Arbitrator shall also have power

- (a) to order that the parties shall agree facts as facts and figures as figures where possible
- (b) to order the parties to prepare an agreed bundle of all documents relevant to the arbitration. The agreed bundle shall thereby be deemed to have been entered in evidence without further proof and without being read out at the Hearing. Provided always that either party may at the Hearing challenge the admissibilty of any document in the agreed bundle.
- (c) to order that any experts whose reports have been exchanged before the Hearing shall be examined by the Arbitrator in the presence of the parties or their legal representatives and not by the parties or their legal representatives themselves. Where such an order is made either party may put questions whether by way of cross-examination or re-examination to any party's expert after all experts have been examined by the Arbitrator provided that the party so doing shall first give notice of the nature of the questions he wishes to put.

13.2   Before the Hearing the Arbitrator may and shall if so requested by the parties read the documents to be used at the Hearing. For this or any other purpose the Arbitrator may require all such documents to be delivered to him at such time and place as he may specify.

Rule 14. Summary Awards

14.1   The Arbitrator may at any time make a *Summary Award* and for this purpose shall have power to award payment by one party to another of a sum representing a reasonable proportion of the final nett amount which in his opinion that party is likely to be

ordered to pay after determination of all the issues in the arbitration and after taking into account any defence or counterclaim upon which the other party may be entitled to rely.
14.2   The Arbitrator shall have power to order the party against whom a Summary Award is made to pay part or all of the sum awarded to a stakeholder. In default of compliance with such an order the Arbitrator may order payment of the whole sum in the Summary Award to the other party.
14.3   The Arbitrator shall have power to order payment of costs in relation to a Summary Award including power to order that such costs shall be paid forthwith.
14.4   A Summary Award shall be final and binding upon the parties unless and until it is varied by any subsequent Award made and published by the same Arbitrator or by any other arbitrator having jurisdiction over the matters in dispute. Any such subsequent Award may order repayment of monies paid in accordance with the Summary Award.

Part D. Procedure at the Hearing

Rule 15. The Hearing
15.1   At or before the Hearing and after hearing representations on behalf of each party the Arbitrator shall determine the order in which the parties shall present their cases and/or the order in which the issues shall be heard and determined.
15.2   The Arbitrator may order any submission or speech by or on behalf of any party to be put into writing and delivered to him and to the other party. A party so ordered shall be entitled if he so wishes to enlarge upon or vary any such submission orally.
15.3   The Arbitrator may on the application of either party or of his own motion hear and determine any issue or issues separately.
15.4   If a party fails to appear at the Hearing and provided that the absent party has had notice of the Hearing or the Arbitrator is satisfied that all reasonable steps have been taken to notify him of the Hearing the Arbitrator may proceed with the Hearing in his absence. The Arbitrator shall nevertheless take all reasonable steps to ensure that the real issues between the parties are determined justly and fairly.

Rule 16. Evidence
16.1   The Arbitrator may order a party to submit in advance of the Hearing a list of the witnesses he intends to call. That party shall not thereby be bound to call any witness so listed and may add to the list so submitted at any time.

16.2   No expert evidence shall be admissible except by leave of the Arbitrator. Leave may be given on such terms and conditions as the Arbitrator thinks fit. Unless the Arbitrator otherwise orders such terms shall be deemed to include a requirement that a report from each expert containing the substance of the evidence to be given shall be served upon the other party within a reasonable time before the Hearing.

16.3   The Arbitrator may order disclosure or exchange or proofs of evidence relating to factual issues. The Arbitrator may also order any party to prepare and disclose in advance a list of points or questions to be put in cross-examination of any witness.

16.4   Where a list of questions is disclosed whether pursuant to an order of the Arbitrator or otherwise the party making disclosure shall not be bound to put any question therein to the witness unless the Arbitrator so orders. Where the party making disclosure puts a question not so listed in cross-examination the Arbitrator may disallow the costs thereby occasioned.

16.5   The Arbitrator may order that any proof of evidence which has been disclosed shall stand as the evidence in chief of the deponent provided that the other party has been or will be given an opportunity to cross-examine the deponent thereon. The Arbitrator may also at any time before such cross-examination order the deponent or some other identified person to deliver written answers to questions arising out of the proof of evidence.

16.6   The Arbitrator may himself put questions to any witness and/or require the parties to conduct enquiries tests or investigations. Subject to his agreement the parties may ask the Arbitrator to conduct or arrange for any enquiry test or investigation.

Part E. After the Hearing

Rule 17. The Award
17.1   Upon the closing of the Hearing (if any) and after having considered all the evidence and submissions the Arbitrator will prepare and publish his Award.

17.2   When the Arbitrator has made and published his Award (including a Summary Award under Rule 14) he will so inform the parties in writing and shall specify how and where it may be taken up upon due payment of his fee.

Rule 18. Reasons
18.1   Whether requested by any party to do so or not the Arbitrator may at his discretion state his Reasons for all or any

part of his Award. Such Reasons may form part of the Award itself or may be contained in a separate document.

18.2   A party asking for Reasons shall state the purpose for his request. If the purpose is to use them for an appeal (whether under S.1 of the Arbitration Act 1979 or otherwise) the requesting party shall also specify the points of law with which he wishes the Reasons to deal. In that event the Arbitrator shall give the other party an opportunity to specify additional points of law to be dealt with.

18.3   Reasons prepared as a separate document may be delivered with the Award or later as the Arbitrator thinks fit.

18.4   Where the Arbitrator decides not to state his Reasons he shall nevertheless keep such notes as will enable him to prepare Reasons later if so ordered by the High Court.

Rule 19. Appeals

19.1   If any party applies to the High Court for leave to appeal against any Award or decision or for an order staying the arbitration proceedings or for any other purpose that party shall forthwith notify the Arbitrator of the application.

19.2   Once any Award or decision has been made and published the Arbitrator shall be under no obligation to make any statement in connection therewith other than in compliance with an order of the High Court under S.1(5) of the Arbitration Act 1979.

Part F. Short Procedure

Rule 20. Short Procedure

20.1   Where the parties so agree (either of their own motion or at the invitation of the Arbitrator) the arbitration shall be conducted in accordance with the following *Short Procedure*.

20.2   Each party shall set out his case in the form of a file containing

(a) a statement as to the orders or awards he seeks
(b) a statement of his reasons for being entitled to such orders or awards and
(c) copies of any documents on which he relies (including statements) identifying the origin and date of each document

and shall deliver copies of the said file to the other party and to the Arbitrator in such manner and within such time as the Arbitrator may direct.

20.3   After reading the parties' cases the Arbitrator may view the site or the Works and may require either or both parties to submit further documents or information in writing.

20.4   Within one calendar month of completing the foregoing steps the Arbitrator shall fix a day when he shall meet the parties for the purpose of

(a) receiving any oral submissions which either party may wish to make and/or

(b) the Arbitrator's putting questions to the parties their representatives or witnesses

For this purpose the Arbitrator shall give notice of any particular person he wishes to question but no person shall be bound to appear before him.

20.5   Within one calendar month following the conclusion of the meeting under Rule 20.4 or such further period as the Arbitrator may reasonably require the Arbitrator shall make and publish his Award.

Rule 21. Other matters

21.1   Unless the parties otherwise agree the Arbitrator shall have no power to award costs to either party and the Arbitrator's own fees and charges shall be paid in equal shares by the parties. Where one party has agreed to the Arbitrator's fees the other party by agreeing to this Short Procedure shall be deemed to have agreed likewise to the Arbitrator's fees.

21.2   Either party may at any time before the Arbitrator has made and pubished his Award under this Short Procedure require by written notice served on the Arbitrator and the other party that the arbitration shall cease to be conducted in accordance with this Short Procedure. Save only for Rule 21.3 the Short Procedure shall thereupon no longer apply or bind the parties but any evidence already laid before the Arbitrator shall be admissible in further proceedings as if it has been submitted as part of those proceedings and without further proof.

21.3   The party giving written notice under Rule 21.2 shall thereupon in any event become liable to pay

(a) the whole of the Arbitrator's fees and charges incurred up to the date of such notice and

(b) a sum to be assessed by the Arbitrator as reasonable compensation for the costs (including any legal costs) incurred by the other party up to the date of such notice.

Payment in full of such charges shall be a condition precedent to that party's proceeding further in the arbitration unless the

Arbitrator otherwise directs. Provided that non-payment of the said charges shall not prevent the other party from proceeding in the arbitration.

Part G. Special Procedure for Experts

Rule 22. Special Procedure for Experts

22.1   Where the parties so agree (either of their own motion or at the invitation of the Arbitrator) the hearing and determination of any issues of fact which depend upon the evidence of experts shall be conducted in accordance with the following *Special Procedure*.

22.2   Each party shall set out his case on such issues in the form of a file containing

(a) a statement of the factual findings he seeks
(b) a report or statement from and signed by each expert upon whom that party relies and
(c) copies of any other documents referred to in each expert's report or statement or on which the party relies identifying the origin and date of each document

and shall deliver copies of the said file to the other party and to the Arbitrator in such manner and within such time as the Arbitrator may direct.

22.3   After reading the parties' cases the Arbitrator may view the site or the Works and may require either or both parties to submit further documents or information in writing.

22.4   Thereafter the Arbitrator shall fix a day when he shall meet the experts whose reports or statements have been submitted. At the meeting each expert may address the Arbitrator and put questions to any other expert representing the other party. The Arbitrator shall so direct the meeting as to ensure that each expert has an adequate opportunity to explain his opinion and to comment upon any opposing opinion. No other person shall be entitled to address the Arbitrator or question any expert unless the parties and the Arbitrator so agree.

22.5   Thereafter the Arbitrator may make and publish an Award setting out with such details or particulars as may be necessary his decision upon the issues dealt with.

Rule 23. Costs

23.1   The Arbitrator may in his Award make orders as to the payment of any costs relating to the foregoing matters including his own fees and charges in connection therewith.

23.2 Unless the parties otherwise agree and so notify the Arbitrator neither party shall be entitled to any costs in respect of legal representation assistance or other legal work relating to the hearing and determination of factual issues by this Special Procedure.

Part H. Interim Arbitration

Rule 24. Interim Arbitration

24.1 Where the Arbitrator is appointed and the arbitration is to proceed before completion or alleged completion of the Works then save in the case of a dispute arising under Clause 63 of the ICE Conditions of Contract the following provisions shall apply in addition to the foregoing Rules and the arbitration shall be called an Interim Arbitration.

24.2 In conducting an Interim Arbitration the Arbitrator shall apply the powers at his disposal with a view to making his Award or Awards as quickly as possible and thereby allowing or facilitating the timely completion of the Works.

24.3 Should an Interim Arbitration not be completed before the Works or the relevant parts thereof are complete the Arbitrator shall within 14 days of the date of such completion make and publish his Award findings of fact or Interim Decision pursuant to Rule 24.5 hereunder on the basis of evidence given and submissions made up to that date together with such further evidence and submissions as he may in his discretion agree to receive during the said 14 days. Provided that before the expiry of the said 14 days the parties may otherwise agree and so notify the Arbitrator.

24.4 For the purpose only of Rule 24.3 the Arbitrator shall decide finally whether and if so when the Works or the relevant parts thereof are complete.

24.5 In an Interim Arbitration the Arbitrator may make and publish any or all of the following

(a) a Final Award or an Interim Award on the matters at issue therein
(b) findings of fact
(c) a Summary Award in accordance with Rule 14
(d) an Interim Decision as defined in Rule 24.6.

An Award under (a) above or a Finding under (b) above shall be final and binding upon the parties in any subsequent proceedings. Anything not expressly identified as falling under either of headings (a) (b) or (c) above shall be deemed to be an Interim Decision under heading (d). Save as aforesaid the Arbitrator shall not make an Interim Decision without first notifying the parties that he intends to do so.

24.6   An *Interim Decision* shall be final and binding upon the parties and upon the Engineer (if any) until such time as the Works have been completed or any Award or decision under Rule 24.3 has been given. Thereafter the Interim Decision may be re-opened by another Arbitrator appointed under these Rules and where such other Arbitrator was also the Arbitrator appointed to conduct the Interim Arbitration he shall not be bound by his earlier Interim Decision.

24.7   The Arbitrator in an Interim Arbitration shall have power to direct that Part F (Short Procedure) and/or Part G (Special Procedure for Experts) shall apply to the Interim Arbitration.

Part J. Miscellaneous

Rule 25. Definitions
25.1   In these Rules the following definitions shall apply.

(a) 'Arbitrator' includes a tribunal of two or more Arbitrators or an Umpire.

(b) 'Institution' means The Institution of Civil Engineers.

(c) 'ICE Conditions of Contract' means the Conditions of Contract for use in connection with Works of Civil Engineering Construction published jointly by the Institution, the Association of Consulting Engineers and the Federation of Civil Engineering Contractors.

(d) 'Other party' includes the plural unless the context otherwise requires.

(e) 'President' means the President for the time being of the Institution or any Vice-president acting on his behalf.

(f) 'Procedure' means The Institution of Civil Engineers' Arbitration Procedure (1983) unless the context otherwise requires.

(g) 'Award', 'Final Award' and 'Interim Award' have the meanings given to those terms in or in connection with the Arbitration Acts 1950 to 1979. 'Summary Award' means an Award made under Rule 14 hereof.

(h) 'Interim Arbitration' means an arbitration in accordance with Part H of these Rules. 'Interim Decision' means a decision as defined in Rule 24.6 hereof.

Rule 26. Application of the ICE Procedure
26.1   This Procedure shall apply to the conduct of the arbitration if

(a) the parties at any time so agree
(b) the President when making an appointment so directs or
(c) the Arbitrator so stipulates at the time of his appointment

Provided that where this Procedure applies by virtue of the Arbitrator's stipulation under (c) above the parties may within 14 days of that appointment agree otherwise in which event the Arbitrator's appointment shall terminate and the parties shall pay his reasonable charges in equal shares.
26.2   This Procedure shall not apply to arbitrations under the law of Scotland for which a separate *ICE Arbitration Procedure (Scotland)* is available.
26.3   Where an arbitration is governed by the law of a country other than England and Wales this Procedure shall apply to the extent that the applicable law permits.

Rule 27. Exclusion of liability
27.1   Neither the Institution nor its servants or agents nor the President shall be liable to any party for any act omission or misconduct in connection with any appointment made or any arbitration conducted under this Procedure.

Appendix 14.2

Current arbitration clauses incorporated into conditions of contract

Institution of Civil Engineers' Conditions of Contract (fifth edition June 1973) (revised January 1979, reprinted January 1986)

Clause 66. (1) *Settlement of Disputes – Arbitration.* If a dispute or difference of any kind whatsoever shall arise between the Employer and the Contractor in connection with or arising out of the Contract or the carrying out of the Works including any dispute as to any decision opinion instruction direction certificate or valuation of the Engineer (whether during the progress of the Works or after their completion and whether before or after the determination abandonment or breach of the Contract) it shall be referred in writing to and be settled by the Engineer who shall state his decision in writing and give notice of the same to the Employer and the Contractor.

(2) *Engineer's Decision – Effect on Contractor and Employer.* Unless the Contract shall have already been determined or abandoned the Contractor shall in every case continue to proceed with the Works with all due diligence and the Contractor and Employer shall both give effect forthwith to every such decision of the Engineer unless and until the same shall be revised by an arbitrator as hereinafter provided. Such decisions shall be final and binding upon the Contractor and the Employer unless and until the dispute or difference has been referred to arbitration as hereinafter provided and an award made and published.

(3) *Arbitration – Time for Engineer's decision.* (a) Where a Certificate of Completion of the whole of the Works has not been isued and

  (i) either the Employer or the Contractor be dissatisfied with any such decision of the Engineer or

  (ii) the Engineer shall fail to give such decision for a period of one calendar month after such referral in writing

then either the Employer or the Contractor may within 3 calendar months after receiving notice of such decision or within 3 calendar

218

months after the expiration of the said period of one month (as the case may be) refer the dispute or difference to the arbitration of a person to be agreed upon by the parties by giving notice to the other party.

(b) where a Certificate of Completion of the whole of the Works has been issued and

(i) either the Employer or the Contractor be dissatisfied with any such decision of the Engineer or
(ii) the Engineer shall fail to give such decision for a period of 3 calendar months after such referral in writing

then either the Employer or the Contractor may within 3 calendar months after receiving notice of such decision or within 3 calendar months after the expiration of the said period of 3 months (as the case may be) refer the dispute or difference to the arbitration of a person to be agreed upon by the parties by giving notice to the other party.

(4) *President or Vice-President to act.* (a) If the parties fail to appoint an arbitrator within one calendar month of either party serving on the other party a written Notice to Concur in the appointment of an arbitrator the dispute or difference shall be referred to a person to be appointed on the application of either party by the President for the time being of the Instutition of Civil Engineers.

(b) If an arbitrator declines the appointment or after appointment is removed by order of a competent court or is incapable of acting or dies and the parties do not within one calendar month of the vacancy arising fill the vacancy then either party may apply to the President for the time being of the Institution of Civil Engineers to appoint another arbitrator to fill the vacancy.

(c) In any case where the President for the time being of the Institution of Civil Engineers is not able to exercise the functions conferred on him by this Clause the said functions may be exercised on his behalf by a Vice-President for the time being of the said Institution.

(5) *ICE Arbitration Procedure (1983).* (a) Any reference to arbitration shall be conducted in accordance with the Institution of Civil Engineers' Arbitration Procedure (1983) or any amendment or modification thereof being in force at the time of the appointment of the arbitrator. Such arbitrator shall have full power to open up review and revise any decision opinion instruction direction certificate or valuation of the Engineer and neither party shall be limited in the proceedings before such

arbitrator to the evidence or arguments put before the Engineer for the purpose of obtaining his decision above referred to.

(b) Any such reference to arbitration shall be deemed to be a submission to arbitration within the meaning of the Arbitration Act 1950 or any statutory re-enactment or amendment thereof for the time being in force. The award of the arbitrator shall be binding on all parties.

(c) Any reference to arbitration may unless the parties otherwise agree in writing proceed notwithstanding that the Works are not then complete or alleged to be complete.

(6) *Engineer as witness.* No decision given by the Engineer in accordance with the foregoing provisions shall disqualify him from being called as a witness and giving evidence before the arbitrator on any matter whatsoever relevant to the dispute or difference so referred to the arbitrator as aforesaid.

Clause 67. Application to Scotland. (1) If the Works are situated in Scotland the Contract shall in all respects be construed and operate as a Scottish contract and shall be interpreted in accordance with Scots Law and the provisions of this Clause shall apply.

(2) In the application of Clause 66 the word 'arbiter' shall be substituted for the word 'arbitrator'. Any reference to the Arbitration Act 1950 shall be deleted and for any reference to the Institution of Civil Engineers' Arbitration Procedure (1983) there shall be substituted a reference to the Institution of Civil Engineers' Arbitration Procedure (Scotland) (1983).

Overseas (Civil) Conditions of Contract
(first edition August 1956)

Clause 66. Settlement of Disputes – Arbitration. If any dispute or difference of any kind whatsoever shall arise between the Employer or the Engineer and the Contractor in connection with or arising out of the Contract or the carrying out of the Works (whether during the progresss of the Works or after their completion and whether before or after the determination abandonment or breach of the Contract) it shall be referred to and settled by the Engineer who shall state his decision in writing and give notice of the same to the Employer and the Contractor. Such decision in respect of every matter so referred shall be final and binding upon the Employer and the Contractor until the completion of the work and shall forthwith be given effect to by the Contractor who shall proceed with the Works with all due diligence whether notice of dissatisfaction is given by him or by the Employer as hereinafter provided or not. If the Engineer shall fail to give such decision for a period of 90 days after being requested to do so or if either the Employer or the Contractor be dissatisfied with any such decision of the Engineer then and in any such case either the Employer or the Contractor may within 90 days after receiving notice of such decision or within 90 days after the expiration of the first named period of 90 days (as the case may be) require that the matter shall be referred to an arbitrator to be agreed upon between the parties or failing agreement to be nominated on the application of either party by the appointor designated in the form of tender for that purpose and any such reference shall be deemed to be a submission to arbitration within the meaning of the Arbitration Laws of the country to the Law of which the Contract is subject but if the Engineer has given a decision and given notice thereof as aforesaid within a period of 90 days as aforesaid and no notice of dissatisfaction has been given either by the Employer or the Contractor within a period of 90 days from receipt of such notice thereof the said decision of the Engineer shall remain final and binding upon the Employer and the Contractor. Such arbitrator shall have full power to open up review and revise any decision opinion direction certificate or valuation of the Engineer and neither party shall be limited in the proceedings before such arbitrator to the evidence or arguments put before the Engineer for the purpose of obtaining his decision above referred to. The award of the arbitrator shall be final and binding on the parties. Such reference except as to the withholding by the Engineer of any certificate or the withhholding of any portion of the retention money under Clause 60 hereof to which the Contractor claims to be entitled or as to the exercise of the Engineer's power to give a certificate under Clause 63 (1) hereof shall not be opened until after the completion or alleged completion of the Works unless with the written consent of the Employer and the Contractor. Provided always:

(i) That the giving of a Certificate of Completion under Clause 48 hereof shall not be a condition precedent to the opening of any such reference
(ii) That no decision given by the Engineer in accordance with the foregoing provisions shall disqualify him from being called as a witness and giving evidence before the arbitrator on any matter whatsoever relevant to the dispute or difference so referred to the arbitrator as aforesaid.

FIDIC Conditions of Contract for Works of Civil Engineering Construction, Part 1 – General Conditions (fourth edition, 1987)

Sub-Clause 67.1 Engineer's Decision. If a dispute of any kind whatsoever arises between the Employer and the Contractor in connection with, or arising out of, the Contract or the execution of the Works, whether during the execution of the Works or after their completion and whether before or after repudiation or other termination of the Contract, including any dispute as to any opinion, instruction, determination, certificate or valuation of the Engineer, the matter in dispute shall, in the first place, be referred in writing to the Engineer, with a copy to the other party. Such reference shall state that it is made pursuant to this Clause. No later than the eighty-fourth day after the day on which he received such reference the Engineer shall give notice of his decision to the Employer and the Contractor. Such decision shall state that it is made pursuant to this Clause.

Unless the Contract has already been repudiated or terminated, the Contractor shall, in every case, continue to proceed with the Works with all due diligence and the Contractor and the Employer shall give effect forthwith to every such decision of the Engineer unless and until the same shall be revised, as hereinafter provided, in an amicable settlement or an arbitral award.

If either the Employer or the Contractor be dissatisfied with any decision of the Engineer, or if the Engineer fails to give notice of his decision on or before the eighty-fourth day after the day on which he received the reference, then either the Employer or the Contractor may, on or before the seventieth day after the day on which he received notice of such decision, or on or before the seventieth day after the day on which the said period of 84 days expired, as the case may be, give notice to the other party, with a copy for information to the Engineer, of his intention to commence arbitration, as hereinafter provided as to the matter in dispute. Such notice shall establish the entitlement of the party giving the same to commence arbitration, as hereinafter provided, as to such dispute and, subject to Sub-Clause 67.4, no arbitration in respect thereof may be commenced unless such notice is given.

If the Engineer has given notice of his decision as to a matter in dispute to the Employer and the Contractor and no notification of intention to commence arbitration as to such dispute has been given by either the Employer or the Contractor on or before the seventieth day after the day on which the parties received notice as to such decision from the Engineer, the said decision shall become final and binding upon the Employer and the Contractor.

Sub-Clause 67.2 Amicable Settlement Where notice of intention to commence arbitration as to a dispute has been given in accordance with Sub-Clause 67.1, arbitration of such dispute shall not be commenced unless an attempt has first been made by the parties to settle such dispute amicably. Provided that, unless the parties otherwise agree, arbitration may be commenced on or after the fifty-sixth day after the day on which notice of intention to commence arbitration of such dispute was given, whether or not any attempt at amicable settlement thereof has been made.

Sub-Clause 67.3 Arbitration. Any dispute in respect of which:

(a) the decision, if any of the Engineer has not become final and binding pursuant to Sub-Clause 67.1, and

(b) amicable settlement has not been reached within the period stated in Sub-Clause 67.2

shall be finally settled, unless otherwise specified in the Contract, under the Rules of Conciliation and Arbitration of the International Chamber of Commerce by one or more arbitrators appointed under such Rules. The said arbitrator/s shall have full power to open up, review and revise any decision, opinion, instruction, determination, certificate or valuation of the Engineer related to the dispute.

Neither party shall be limited in the proceedings before such arbitrator/s to the evidence or arguments put before the Engineer for the purpose of obtaining his said decision pursuant to Sub-Clause 67.1. No such decision shall disqualify the Engineer from being called as a witness and giving evidence before the arbitrator/s on any matter whatsoever relevant to the dispute.

Arbitration may be commenced prior to or after completion of the Works, provided that the obligations of the Employer, the Engineer and the Contractor shall not be altered by reason of the arbitration being conducted during the progress of the Works.

Sub-Clause 67.4 Failure to Comply with Engineer's Decision. Where neither the Employer nor the Contractor has given notice of intention to commence arbitration of a dispute within the period stated in Sub-Clause 67.1 and the related decision has become final and binding, either party may, if the other party fails to comply with such decision, and without prejudice to any other rights it may have, refer the failure to arbitration in accordance with Sub-Clause 67.3. The provisions of Sub-Clauses 67.1 and 67.2 shall not apply to any such reference.

References and further reading

1. *Britain 1988, An Official Handbook*, Central Office of Information, London (1988)
2. *The World Bank and International Finance Corporation*, World Bank, Washington, DC (1986)
3. *The World Bank*, World Bank, Washington, DC (1987)
4. *The World Bank Annual Report 1987*, World Bank, Washington, DC (1987)
5. *Asian Development Bank – Questions and Answers*, Asian Development Bank, Manila (1983)
6. *The OPEC Fund for International Development*, OPEC Fund, Vienna (1987)
7. Smith, J. and Keenan, D., *English Law*, 8th edition, Pitman, London (1986)
8. Wallace, D., *Hudson's Building and Engineering Contracts*, 10th edition, Sweet & Maxwell, London (1970)
9. Keating, D., *Building Contracts*, 4th edition, Sweet & Maxwell, London (1978) with 1982–1984 Supplements
10. Uff, J., *Construction Law*, 4th edition, Sweet & Maxwell, London (1985)
11. Abrahamson, M. W., *Engineering Law: the ICE Contracts*, 4th edition, Applied Science, London (1979)
12. Institution of Civil Engineers, *Civil Engineering Procedure*, 4th edition, Thomas Telford, London (1986)
13. Institution of Civil Engineers, *et al.*, *Conditions of Contract for Use in Connection with Works of Civil Engineering Construction*, 5th edition, ICE, London (1973) (revised 1979, reprinted 1986)
14. Fédération Internationale des Ingénieurs-Conseils (FIDIC), *Conditions of Contract for Works of Civil Engineering Construction*, 4th edition, FIDIC, Lausanne (1987)
15. Federation of Civil Engineering Contractors, *Form of Sub-Contract Designed for Use with ICE Conditions of Contract*, 5th edition, 1973, FCEC, London (1984)
16. *Overseas (Civil) Conditions of Contract*, Association of Consulting Engineers and the Export Group of Constructional Industries (1956)
17. *Institution of Civil Engineers' Conditions of Contract for Ground Investigation*, Thomas Telford, London (1983)
18. *Institution of Civil Engineers' Conditions of Contract for Minor Works*, Thomas Telford, London (1988)
19. *General Conditions of Government Contracts for Building and Civil Enginering Works (Form GC/Works/1)*, 2nd edition, HMSO, London (1977)
20. *General Conditions of Government Contracts for Building and Civil Engineering Minor Works (Form GC/Works/2)*, 2nd edition, HMSO, London (1980)
21. Joint Contracts Tribunal, *Standard Form of Building Contract*, RIBA Publications, London (1980)

22. *Civil Engineering Specification for the Water Industry*, 2nd edition, Water Research Centre and Water Authorities Association (1984)
23. *Department of Transport Specification for Highway Works, Parts 1 to 7*, HMSO, London (1986)
24. Institution of Civil Engineers, *Specification for Piling*, ICE, London (1988)
25. *Civil Engineering Standard Method of Measurement*, 2nd edition, ICE, London (1985)
26. Barnes, M., *The CESMM2 Handbook*, Thomas Telford, London (1986)
27. Seeley, I. H., *Civil Engineering Quantities*, 4th edition, Macmillan, London (1987)
28. Federation of Civil Engineering Contractors, *Schedules of Dayworks Carried Out Incidental to Contract Work*, FCEC, London (1983)
29. *Working Rule Agreement of the Civil Engineering Construction Concilitation Board of Great Britain*, CECCB (1988)
30. Association of Consulting Engineers, *Conditions of Engagement*, ACE, London (1963, 1970 and 1981)
31. Haswell, C. K., 'Rate Fixing in Civil Engineering Contracts', *Proc. ICE* (February 1963)
32. Geddes, S., in G. Chrystal Smith and R. Jolly (eds), *Estimating for Building and Civil Engineering Works*, 8th edition, Butterworths, London (1985)
33. Wood, R. D., *Building and Civil Engineering Claims*, 2nd edition, The Estates Gazette Ltd, London (1978)
34. Walton, A. and Vitoria, M., *Russell on the Law of Arbitration*, 20th edition, Stevens, London (1982)
35. *Arbitration Acts of 1950, 1975 and 1979*, HMSO, London
36. Hawker, G., Uff, J. and Timms, C., *The Institution of Civil Engineers' Arbitration Practice*, Thomas Telford, London (1986)
37. *Rules of the London Court of International Arbitration*, LCIA, London (1985)
38. International Chamber of Commerce, *Rules of Conciliation and Arbitration 1988*, ICC Publishing SA, Paris
39. Ivamy, E. R. H., *General Principles of Insurance Law*, 5th edition, Butterworths, London (1986)
40. Eaglestone, F. N., and Smyth, C., *Insurance for the Construction Industry*, George Godwin, London (1985)

Index

234   Index